ANATOMY AND PHYSIOLOGY

On-The-Go Study Guide to Learning
The Human Anatomy

DARRELL CONNOLLY

TABLE OF CONTENTS

INTRODUCTION

Welcome to the study guide that will change your life.

Learning new things and remembering everything you've learned can be extremely challenging sometimes. It's even harder when you must remember all the concepts in a subject such as Anatomy and Physiology. The problem with this subject is that there are many terms to learn, understand, and remember. Who has the time to sit down and learn everything in this busy, fast-paced world?

The good news is that learning Anatomy and Physiology isn't an impossible task!

When professors are asked why they think students have difficulty learning Anatomy and Physiology, they say there are three possible reasons.

First is the nature of the discipline, referring to the characteristics of the subject itself, how the subject is related to other science subjects, how experts study the subject, how the expert's reason about the subject, and how the experts communicate the knowledge and understanding they have of the discipline.

The second is how the subject is taught, referring s to all the standard practices students encounter in class as the teachers try to help them learn everything about the subject.

Third is what the students bring to their task of learning the subject, referring to the existing skills and knowledge the students already have, how they understand the process of learning, and any "student skills" they've acquired from the time they started studying.

Indicating that Anatomy and Physiology *is* a complex subject to tackle, and you need to put a lot of effort and focus into it. This subject involves loads of terminology, some entirely new and unfamiliar. Of course, if you've already taken a medical terminology course or a few medical subjects in the past, you would be at an advantage. If not, you have a lot to learn! For nursing students knowing these terms is essential, especially when you have to communicate with doctors (so you know what they mean) and patients (so you can explain what the terms mean as needed). As you can see, this subject is complex and challenging, as there is just so much information to learn!

Fortunately, you have this study guide.

The easiest solution to this problem is to take all your study materials with you no matter where you go. So, whenever you have free time, you can use this to review. You need comprehensive, easy-to-understand material containing all the most basic and important terms related to Anatomy and Physiology. In other words, what you need is this study guide.

What's great about study guides is that they contain the information you need but delivered in a synthesized and summarized manner. You can think of study guides as a mini outline, as they only have the most relevant information. Study guides are handy for complex or challenging subjects such as Anatomy and Physiology. One of the primary purposes of study guides is to reduce the amount of information you can learn without compromising the quality of your learning.

Also, as you will see in this study guide, later, you will be able to memorize the terminology more effectively, as presented in an organized and easy-to-understand way. Therefore, you have made the right choice with this guidebook, as it also comes with simple yet effective study tips to help you with your review. You can apply these tips while you study the terminologies written in the following few chapters, and you can also use the tips for your other study sessions.

Does all this sound too good to be true? You might be wondering why you should believe what's written here. After all, you would use this study guide to learn or review the most important terms for a whole course or subject. Your success lies in how practical this book is. To help ease any doubts or reluctance, let me tell you more about myself and why I came up with this book in the first place.

My name is Darrell Connolly, and I am a graduate of Nursing. I am now a Registered Nurse, and I am proud of it. But the road to where I am now wasn't an easy one. After my second year as a nursing student, I felt there was too much to learn, study, and memorize. I was trying to understand and learn everything from all my subjects and the rest of my peers. I struggled with Anatomy and Physiology because this subject involved many medical terms. Since I needed something to help me memorize all these terms, I created this simple reference guide to help me learn all the terms even while I was on the go. It worked so well for me that I decided to have it published for anyone who needs an easy solution to a seemingly impossible problem. After going through this reference guide, you will have the knowledge and confidence to take your exam or use it in the medical field. By using this book, you'll sound like a medical professional in no time.

This study guide is the perfect tool for you to capture the information you need to review for tests or to brush up on your knowledge for when you will start working in real-life medical settings. Let's look at some of the benefits you will enjoy by using this comprehensive guide:

It will allow you to comprehend and reflect on the terms

When you read textbooks, you may encounter a lot of terms used in sentences and examples. Although this may give you a good understanding of the terms, having them presented in outline form allows you to focus on each term individually, comprehend the terms better, and reflect on what they mean. This study guide is more conducive to absorbing the information you need. As you read each of the terms and their definitions, you can visualize them in your mind, read them out loud, and remember them better.

It will prevent you from feeling overwhelmed.

Textbooks are excellent, but they contain SO MUCH information. Think about it - if you had to read an entire text to learn important terms and what they mean, wouldn't this make you feel overwhelmed? Also, trying to learn this way might place a lot of pressure on yourself, making you feel stressed. And when you're stressed, you won't be able to learn effectively. But with this guide, the terms are divided into chapters, making it easier for you to understand them. For instance, Chapter one only contains Anatomical terms. So when you go through them, you can even relate the terms to each other to understand them better, making it much easier than going through paragraphs to find the terms you need and learn what they mean.

It will serve as your permanent reference.

Whether you've already learned these terms before or this is your first time encountering these terms, having this study guide with you means that you have a permanent reference that you can always go through as needed. This study guide was designed to be an on-the-go tool you can bring wherever you go. So, if you have forgotten something, take this book out and look for the proper term. You can also use this guide to help reinforce your learning anytime you need it.

These are some of the most fundamental advantages of using this study guide. The staff I work with at the hospital whom I have given a copy of this book have thanked me several times for helping them learn and memorize these medical terms and their definitions. A lot of them have told me how using this book made their lives so much easier. As I've said, this book doesn't just contain terms and definitions, but I have also shared some highly effective study and memorization tips to help make things easier for you. The subject of Anatomy and Physiology is already a challenge but learning the terms doesn't have to be hard.

With that said, let me give you a promise.

By the end of this short but thorough study guide, you will know the Anatomy and Physiology terms inside and out. You will have a deeper understanding of these terms, and you will be able to use them in your communication because you know what they mean. After going through the chapters in this study guide, you will be able to reinforce the things you have learned in the past and learn new terms you might have never encountered. In this book, you will learn Anatomical, Physiological, and Skeletal terms and the most common prefixes and suffixes used in medical terminology. All this information is presented in an easy-to-understand manner, so you don't have to feel like you're struggling.

So, what are you waiting for?

If you want to jump-start your learning of Anatomy and Physiology, there's no time like the present! I suggest downloading the audiobook version of this book as an audio guide to go along with the book to help with memory retention. Sometimes hearing the terms a few times can help.

Learning these terms allows you to obtain the knowledge you need for examinations or real-life applications. That way, you won't have to worry about having this knowledge later. So, stop wasting your precious time and start learning now! You can download a copy of this book to read it on your device directly during your free time. Or, you can even get the audio version of this study guide and play it while driving or doing chores. Trust me, having this book with you will make a massive difference in your life.

CHAPTER 1

Anatomy Terms

Anatomy might be challenging, but this study guide will make things easier for you. When it comes to anatomy, there are so many terms to learn, all of which are essential to understand the subject better and everything it entails. In this chapter, we'll start with a few study tips which you can use to remember everything you will learn in this guide before moving on to the actual terminology, definitions, and some examples of how the terms are used. Later, we will go through more clever strategies to make learning much easier for you, so you better stick around until the end!

Anatomy is one of the basic subjects which needs to be covered by healthcare students and beyond. Students of nursing, medicine, chiropractic, and osteopathy must learn these terms. Pilates and yoga instructors, massage therapists, fitness trainers, and other health and fitness professionals must also be familiar with these terms to do their jobs better. Healthcare providers and anatomists use these terms, confusing those who haven't studied them. But, the purpose of these terms isn't to confuse people - they are meant to improve precision and avoid medical errors.

When medical professionals use these anatomical terms, this eliminates ambiguity.

Most anatomical terms are derived from Latin and Greek words. Although these languages are now rarely used in daily conversation, their definitions don't change. But before we go into the terms, let's discuss a few study tips to help you.

HELPFUL STUDY TIPS!

Anatomy is the study that focuses on the structure of living organisms. Anatomy is one of the sub-disciplines of biology, further categorized into gross anatomy, the study of large-scale anatomical structures, and microscopic anatomy, the study of microscopic anatomical structures. Human anatomy, which is the focus of this study guide, deals with the anatomical structures of the human body, including organ systems, organs, tissues, and cells. Anatomy is often linked to physiology, which we will discuss in Chapter two. By learning both subjects, you will be able to identify the structure of the human body while getting a better understanding of its functions.

Studying human anatomy gives you a deeper understanding of the body's structures and how these structures work. When you take an introductory course in human anatomy, the goal is to learn and understand the structure and functions of all the body's major systems. However, it would help if you also kept in mind that these organ systems don't exist or work independently. Each system depends on the others, either indirectly or directly, to maintain the body's normal functioning. While learning human anatomy, you will also learn how to identify the other parts of the body, such as the organs, cells, and tissues.

Now that you have a background in the subject, it's time to go through some study tips to help make your learning process more effective. If you keep these tips in mind, you'll find that memorizing all these terms and their definitions becomes a whole lot easier!

For each study or review session, make the most of your time.

When learning or reviewing anatomy, prepare for a lot of memorizations. For example - the human body possesses more than 200 bones and 600 muscles. When you need to learn all these structures, this requires a lot of effort, time, and decent memorization skills. Therefore, each time you have a study or review session, make sure you make the most of your time. You don't have to set long study periods. Instead, it's better to focus on the quality of your study sessions to ensure that you learn and remember new concepts every time.

Familiarize yourself with the language.

The terminology is the most important thing you must understand when studying anatomy. The good news is that this is what you'll learn in this guide. When you're familiar with these terms and know what they mean, you gain a better understanding of the subject and help avoid confusion when it's time for you to work in real-life situations. Because when another medical professional speaks to you about a patient or a condition, you will know precisely what they mean.

Knowing the common prefixes and suffixes is essential, too.

Many medical terms have affixes that also serve as clues to what the terms mean. So if you are already familiar with the terms and the most common prefixes and suffixes, you won't have trouble knowing the functions or definitions of the words you read or hear. More good news for you because we will also go through these common affixes in Chapter four.

Use effective study aids

You will be able to learn anatomy more effectively when you use effective study aids - such as this study guide! These aids can help you gain proficiency in anatomy and all other related subjects. For example, a study guide that presents terminology in an easy-to-understand and organized way makes the learning process smoother and more manageable.

Keep reviewing whenever you have free time.

To ensure that you absorb and comprehend all the information, you must review these terms constantly until you've mastered them. Then, if you're still studying, attend all the anatomy classes and review sessions at school. After reviewing each chapter, you may want to take practice tests and quizzes to assess your learning.

Seek out information.

Finally, you should also ensure that you don't fall behind in information. Anatomy and physiology courses already involve a lot of information, but scientists and researchers are still studying the subject. So, over time, information may get added to the topic, or some info might even change because of new findings. Therefore, after learning the basics, you must continue seeking information to strengthen your learning and deepen your understanding.

Also, if you purchased the physical or eBook version of this study guide, I strongly encourage you to download the audiobook. You can find the link for this audiobook below. On the other hand, if you're listening to this book right now, it would also be a good idea to grab a physical copy or download the eBook to serve as your visual reference guide.

ANATOMY TERMS

Most anatomical terms consist of root words and affixes. Often, the root of these terms refers to a condition, organ, or tissue, whereas the suffix or prefix describes them. For instance, in the medical disorder known as "hypertension," the root here is "tension," which, in this case, means pressure, while the prefix here is "hyper-," which means "over" or "high." Therefore, the term "hypertension" means abnormally high blood pressure. In anatomy, the terms are grouped into different categories. Let's go through these categories one at a time:

Anatomical Position Terms

Anatomists have devised a standardized way to view the human body to increase precision further. Just as explorers would use maps with a normal orientation wherein the north is the top, the standard orientation for anatomical positions is that of the human body standing upright. In this position, the upper limbs are stretched out to either side, the palm of each hand faces forward, the feet are shoulder-width apart, and the toes are facing forward. This standardized position of the human body helps reduce confusion and is known as the "anatomical position." So, no matter what the body's orientation you describe, the terms you would use would refer to this anatomical position. For anatomical positions, the main terms to remember are:

- **Prone** describes a body's face-down orientation. When examining the spine, a patient needs to be prone.

  Prone refers to when the body is lying face-down. Doctors sometimes operate on patients in prone when the area being operated on is at the back.

  Prone is when the human body is in a face-down orientation. To examine the back of the head, the patient must be prone.

- **Supine** describes a body's face-up orientation. Most patients lie supine on hospital beds.

Supine refers to when the body is lying face-up. It's easier to sleep on a hospital bed when supine.

Supine is when the human body is in a face-up orientation. Heart surgeries require patients to be supine.

REGIONAL TERMS

The human body has several regions. Again, to increase precision, specific regional terms are used to describe these body parts.

- **Abdominal region** refers to the stomach. Pain in the **abdominal region** is a symptom of various conditions.

  Abdominal region is where the stomach is located. A lot of people focus on strengthening their **abdominal region** when working out.

  Abdominal region refers to the region of the stomach. Digestion happens in the **abdominal region**.

- **Acromial region** encompasses the shoulders. The scapula's outer end is located in the **acromial region**.

  Acromial region is where the shoulders are located. The **acromial region** is right below the neck.

  Acromial region refers to the shoulders. A dislocated shoulder affects the **acromial region**.

- **Antebrachial region** refers to the lower arm. The **antebrachial region** is one of the more fragile parts of the body.

Antebrachial region is also known as the forearm and is only used to when talking about the lower arm. The **antebrachial region** is connected to the hand.

Antebrachial region is a term used when referring to the lower arm. The **antebrachial region** is found below the upper arm.

- **Antecubital region** encompasses the front part of the elbows. The **antecubital region** contains the antecubital fossa.

Antecubital region is where the front part of the elbow is located. Medical professionals usually draw blood from veins in the **antecubital region**.

Antecubital region refers to the region of the front part of the elbows. The **antecubital region** is anterior to the elbow.

- **Axillary region** encompasses the armpits. The **axillary region** is prone to sweating.

Axillary region is where the armpits are located. The overall shape of the axilla within the **axillary region** is a pyramid.

Axillary region refers to the region of the armpits. The **axillary region** is in the upper chest.

- **Brachial region** refers to the upper arm. Most vaccines are administered in the **brachial region**.

Brachial region is also known as arm and is only used to when talking about the upper arm. The **brachial region** is connected to the shoulder.

Brachial region is a term used when referring to the upper arm. The triceps brachii muscle is the **brachial region's** main muscle mass.

- **Buccal region** refers to the cheeks. The **buccal region** is quite tender.

  Buccal region is where the cheeks are located. The flesh of the **buccal region** is soft and sensitive.

  Buccal region describes the region of the cheeks. The **buccal region** is found on the face.

- **Carpal region** encompasses the wrists. The **carpal region** is proximal to the antebrachial region.

  Carpal region is where the wrists are located. The wrist in the **carpal region** consists of eight bones.

  Carpal region refers to the region of the wrists. A condition called carpal tunnel syndrome mainly affects the **carpal region**.

- **Cephalic region** encompasses the whole head and all its regions. The **cephalic region** can be further divided into smaller regions.

  Cephalic region is where all the regions of the head are located. The **cephalic region** is the topmost part of the body.

  Cephalic region refers to all of the regions of the head. The **cephalic region** is one of the body's major divisions.

- **Cervical region** refers to the neck. The **cervical region** is where the jugular is located.

  Cervical region is where the neck is located. Swallowing is most obvious in the **cervical region**.

  Cervical region refers to the region of the neck. The **cervical region** suffers damage during strangulation.

- **Coxal region** refers to the beltline. The **coxal region** is where the hip bone is located.

  Coxal region is where the beltline is located. The **coxal region** contains the coxal gland.

  Coxal region refers to the region of the beltline. One of the bones in the **coxal region** is known as the ischium.

- **Cranial region** refers to the upper portion of the head. His injury posed a threat to his brain as it was in his **cranial region**.

  Cranial region describes the head's upper portion. It's important to prevent the **cranial region** from being hit with blunt force.

  Cranial region is located on the upper part of the head. The doctor had to operate on his **cranial region** to remove a tumor.

- **Crural region** encompasses the lower legs. The **crural region** is located below the abdominal region.

Crural region is where the lower legs are located. The tibia is in the **crural region**.

Crural region refers to the region of the lower legs. The body has posterior and anterior **crural regions**.

- **Crus** refers to the part of our lower limb between the ankle and knee. It's extremely painful to be kicked in the **crus**.

  Crus is the lower limb found between the knee and ankle. The **crus** is directly connected to the ankle.

  Crus is a part of the lower limb located between the ankle and knee. The **crus** is directly connected to the knee.

- **Dorsal region** encompasses the upper back. The dorsal body cavity is in the **dorsal region**.

  Dorsal region is where the upper part of the back is located. Part of the spine is in the **dorsal region**.

  Dorsal region refers to the upper part of the back. The **dorsal region** is located above the lumbar region.

- **Facial region** refers to the lower portion of the head starting beneath the ears. Your **facial region** starts under your ears.

  Facial region describes the head's lower portion from below the ears. The jaw is part of the **facial region**.

  Facial region is located on the lower part of the head beginning under the ears. The **facial region** is connected to the neck.

- **Femoral region** encompasses the thighs. The **femoral region** contains a lot of veins and arteries.

  Femoral region is where the thighs are located. The **femoral region** is one of the most common regions associated with buboes.

  Femoral region refers to the region of the thighs. Twisted nerves in the **femoral region** may cause numbness in the thighs.

- **Gluteal region** encompasses the buttocks. The **gluteal region** is at the posterior of the pelvic girdle.

  Gluteal region is where the buttocks is located. The gluteus maximus is the largest muscle in the **gluteal region**.

  Gluteal region refers to the buttocks. Other major muscles of the **gluteal region** are the gluteus minimus and gluteus medius.

- **Groin region** refers to the area between the genitals and legs. The **groin region** is also known as the inguinal region.

  Groin region describes the area between the legs and genitals. The **groin region** contains a gland which secretes a musky odor.

  Groin region refers to the region found between the genitals and the legs. The **groin region** is susceptible to inguinal hernias.

- **Lower limb region** encompasses all of the leg's regions. The **lower limb region** includes all of the lower extremities.

Lower limb region is where all the regions of the legs are located. The **lower limb region** can be further divided into smaller regions.

Lower limb region refers to all the regions of the legs. There are a lot of bones and muscles in the **lower limb region**.

- **Lumbar region** encompasses the lower back. The **lumbar region** lies lateral to the lumbar vertebrae.

  Lumbar region is where the lower part of the back is located. Sometimes, the **lumbar region** is also known as the lower spine.

  Lumbar region refers to the lower part of the back. Three protective layers of tissue in the **lumbar region** surround the spinal cord.

- **Mammary region** refers to both breasts. Men and women have different **mammary regions**.

  Mammary region is where both breasts are located. The **mammary region** of females produces milk for infants.

  Mammary region refers to the region of the breasts. It's important to have our **mammary region** checked regularly.

- **Manual region** encompasses the backs of the hands. The **manual region** is also known as the manus region.

  Manual region is where the backs of the hands are located. The **manual region** is one of the hand regions.

Manual region refers to the backs of the hands. The **manual region** is the posterior region of the hand.

- **Mental region** refers to the chin. People have varying **mental regions** in terms of shapes and sizes.

  Mental region is where the chin is located. The **mental region** is found below the mouth.

  Mental region refers to the region of the chin. The **mental region** is located between the jaws.

- **Nasal region** refers to the nose. Our **nasal region** can get clogged up in cold weather.

  Nasal region is where the nose is located. It's important to keep your **nasal region** clear.

  Nasal region refers to the region of the nose. We breathe mainly through our **nasal region**.

- **Ocular region** refers to the eyes. It's important to take very good care of your **ocular region**.

  Ocular region is where the eyes are located. The **ocular region** is very sensitive.

  Ocular region refers to the region of the eyes. Injuries to the **ocular region** can be very dangerous.

- **Olecranal region** encompasses the back part of the elbows. The olecranon, which is part of the ulna, is in the **olecranal region**.

Olecranal region is where the back part of the elbows is located. The **olecranal region** only bends one way.

Olecranal region refers to the back part of the elbows. The **olecranal region** is found in the arms.

- **Otic region** refers to the ears. It's important to maintain the cleanliness of the **otic region**.

  Otic region is also known as the auricle where the ears are located. You may cover your **otic region** to protect it from loud sounds.

  Otic region refers to the region of the ears. The **otic region** is a very sensitive part of the body.

- **Palmar region** encompasses the palms. The **palmar region** is another part of the body that's prone to sweating.

  Palmar region is where the palms are located. Wearing gloves covers the **palmar region**.

  Palmar region refers to the region of the palms. One of the more complex regions of the human body is the **palmar region**.

- **Patellar region** encompasses the knees. The **patellar region** is in front of the popliteal surface region.

  Patellar region is where the knees are located. The knees in the **patellar region** bend when we walk.

  Patellar region refers to the region of the knees. The kneecap or "patella" is found in the **patellar region**.

- **Pedal region** encompasses the feet. We use our **pedal region** when we walk.

  Pedal region is where the feet are located. The **pedal region** develops and strengthens as children grow older.

  Pedal region refers to the region of the feet. The **pedal region** is important, especially for runners.

- **Phalangeal region** encompasses the toes or fingers. The **phalangeal region** is also known as the digital region.

  Phalangeal region is where the toes or fingers are located. The **phalangeal region** is particularly susceptible to frostbite.

  Phalangeal region refers to the region of the fingers or toes. When referring to the **phalangeal region**, this can either be about the toes or fingers.

- **Plantar region** encompasses the soles of the feet. The **plantar region** is the bottom-most part of the body.

  Plantar region is where the soles of the feet are located. The **plantar region** is also known as the plantar aspect.

  Plantar region refers to the soles of the feet. The **plantar region** is essential for walking, running, standing, and more.

- **Popliteus region** encompasses the back part of the knees. The popliteus is one of the muscles in the **popliteus region**.

Popliteus region is where the back part of the knees is located. The muscles in the **popliteus region** are important for walking.

Popliteus region refers to the back part of the knees. The popliteal fossa in the **popliteus region** is also known as the kneepit.

- **Pubic region** refers to the area located above the genitals. Both males and females have a **pubic region**.

  Pubic region encompasses the area around the genitals. The **pubic region** includes the navel or umbilicus.

  Pubic region refers to the region surrounding the genitals. You can find the **pubic region** in the lower part of the abdominal region.

- **Sural region** encompasses the back part of the lower legs. The **sural region** is the muscular swelling located at the back of the legs.

  Sural region is where the back part of the lower legs is located. The major muscles in the **sural region** are the soleus and gastrocnemius muscles.

  Sural region refers to the back part of the lower legs. The **sural region** also includes the calves.

- **Tarsal region** encompasses the ankles. The calcaneus is the largest tarsal bone in the **tarsal region** of human beings.

  Tarsal region is where the ankles are located. Tarsal tunnel syndrome is a condition which affects the **tarsal region**.

arsal region refers to the region of the ankles. The tarsals are a set of bones in the **tarsal region**.

- **Thoracic region** refers to the chest. The **thoracic region** is where the heart is located.

  Thoracic region is where the chest is located. Injuries to the **thoracic region** can prove fatal.

  Thoracic region refers to the region of the chest. The lungs are in the **thoracic region**.

- **Upper limb region** encompasses all the arm's regions. The **upper limb region** includes all the upper extremities.

  Upper limb region is where all the regions of the arms are located. The **upper limb region** can be further divided into smaller regions.

  Upper limb region refers to all the regions of the arms. There are a lot of bones and muscles in the **upper limb region**.

DIRECTIONAL TERMS

Anatomy also involves several directional terms to describe the relative locations and directions of the body's various structures. It's important to remember all these terms to avoid any confusion, whether you're still studying, or you need to work with other medical professionals.

- **Anterior** refers to the front of the body. Our toes are **anterior** to our feet.

Anterior describes the direction toward the front of the body. The word **anterior** is also known as ventral.

Anterior refers to the direction toward the body's front. A woman's breasts are **anterior** to her shoulder blades.

- **Deep** refers to a position that's far from the body's surface. Our brain is **deep** to our skull.

Deep describes a position that's farther from the surface of the body. The spine is **deep** to the skin.

Deep refers to a position that's far from the surface of the body. The bones are **deep** to the muscles.

- **Distal** refers to a position of a limb that's far from the attachment point. The foot is **distal** to the knee.

Distal describes a body part's position that's farther to the point of attachment. The hand is **distal** to the elbow.

Distal refers to a position of a limb that's far from the body's trunk. The toes are **distal** to the ankle.

- **Inferior** refers to a position lower than another body part. The groin is **inferior** to the abdomen.

Inferior describes a position below another body part. The word **inferior** is also known as caudal.

Inferior refers to a position lower than another part of the body. The feet are **inferior** to the knees.

- **Lateral** refers to the side of the body. Our thumbs are **lateral** to the rest of the digits.

Lateral describes the direction toward the side of the body. The shoulders are **lateral** to the neck.

Lateral refers to the direction toward the body's side. Our smallest toes are **lateral** to the rest of our toes.

- **Medial** refers to the middle of the body. The belly button is the **medial** of the stomach.

Medial describes the direction toward the middle of the body. The **medial** toe is known as the hallux.

Medial refers to the direction toward the body's middle. The **medial** finger is known as the digitus medius.

- **Posterior** refers to the back of the body. Our nape is **posterior** to our jaw.

Posterior describes the direction toward the back of the body. The word **posterior** is also known as dorsal.

Posterior refers to the direction toward the body's back. The spine is **posterior** to the chest.

- **Proximal** refers to a position of a limb that's near the attachment point. The shoulder is **proximal** to the upper arm.

Proximal describes a body part's position that's close to the point of attachment. The ankle is **proximal** to the foot.

Proximal refers to a position of a limb that's near the body's trunk. The thigh is **proximal** to the lower leg.

- **Superior** refers to a position higher than another body part. The head is **superior** to the feet.

  Superior describes a position above another body part. The word **superior** is also known as cranial.

  Superior refers to a position higher than another part of the body. The neck is **superior** to the chest.

BODY PLANES

The term "plane" refers to an imaginary 2D surface that goes right through the human body. In medicine and anatomy, there are three major planes to remember:

- **Frontal plane** refers to the plane dividing an organ into anterior and posterior portions. Adduction along the body's **frontal plane** causes the arm to move in a curvilinear arc.

  Frontal plane refers to the plane dividing the body into front and rear portions. It's possible for exact alignment to occur in the **frontal plane** when three precise points are collinear.

  Frontal plane refers to the plane dividing the whole body or one of the organs into front (anterior) and rear (posterior) portions. To make orientation easier, consider the thumbnail resting on the **frontal plane** of the thumb.

- **Sagittal plane** refers to the plane dividing an organ vertically into left and right sides. In a **sagittal plane**, you can flex, extend or hyper-extend your neck.

Sagittal plane refers to the plane dividing the body vertically into left and right sides. In a **sagittal plane**, reaching away from your body lengthens your spine.

Sagittal plane refers to the plane dividing the whole body or one of the organs vertically into left and right sides. There are times when a basic arthrodesis procedure won't correct a significant imbalance in the **sagittal plane**.

- **Transverse plane** refers to the plane dividing an organ horizontally into lower and upper parts. An image produced by a **transverse plane** is called a cross-section.

  Transverse plane refers to the plane dividing the body horizontally into lower and upper parts. Most plasma responses only occur in a **transverse plane**.

  Transverse plane refers to the plane dividing the whole body or one of the organs horizontally into lower and upper parts. In most patients, **transverse plane** images provide optimal identification of the mechanism and location of stenosis.

BODY CAVITIES

The primary purpose of the body cavities is to contain the internal organs to protect them. The main body cavities are:

- **Abdominal cavity** refers to the cavity that houses the abdominal organs. Sometimes, the **abdominal cavity** may get distended with carbon dioxide.

Abdominal cavity refers to the cavity that contains the intestines, stomach, gallbladder, spleen, kidneys, ureters, and pancreas. There are conditions which cause blood to accumulate in the **abdominal cavity**.

Abdominal cavity refers to the cavity that houses the abdominal organs which are the intestines, stomach, gallbladder, spleen, kidneys, ureters, and pancreas. Hematoceles may occur in the **abdominal cavity**.

- **Cranial cavity** refers to a large cavity within the skull. When the brain swells in a closed **cranial cavity**, this can be very serious.

Cranial cavity refers to a cavity with a bean shape that contains the brain. Mature brain tissue known as glial choristomas can be found outside of the **cranial cavity**.

Cranial cavity refers to a large cavity with a bean shape where the brain is found. The **cranial cavity** has walls which are smooth and thin.

- **Pelvic cavity** refers to the cavity that houses the pelvic organs. There are times when fluid builds up in the **pelvic cavity**.

Pelvic cavity refers to the cavity that contains the urethra, urinary bladder, the rectum, uterus, part of the large intestine, prostate in males, and vagina in females. The ovaries are located right along the lateral walls of the **pelvic cavity**.

Pelvic cavity refers to the cavity that houses the pelvic organs which are the urethra, urinary bladder, the rectum, uterus, part of the large intestine, prostate in males, and vagina in females. In rare cases, ependymoma can occur in the **pelvic cavity**.

- **Spinal cavity** refers to the cavity consisting of the spinal column. Injuries to the **spinal cavity** are very dangerous.

    Spinal cavity refers to the cavity that contains the spinal column and is connected to the cranial cavity. The **spinal cavity** is part of the dorsal cavity.

    Spinal cavity refers to the cavity consisting of the spinal column that's connected to the cranial cavity. The **spinal cavity** is directly connected to the cranial cavity.

- **Thoracic cavity** refers to the cavity that contains the organs in the upper chest. The **thoracic cavity** is also known as the chest cavity.

    Thoracic cavity refers to the cavity that contains the trachea, lungs, esophagus, aorta, and heart. The **thoracic cavity** can be further divided into smaller cavities which are the pleural cavity and the mediastinum.

    Thoracic cavity refers to the cavity that contains the organs in the upper chest which are the trachea, lungs, esophagus, aorta, and heart. The **thoracic cavity** contains some of the body's most important organs.

CHAPTER 2

Physiology Terms

Physiology refers to the study of how living organisms' function. In the word physiology, the term "physi" comes from the Greek word that broadly translates to "natural origin." This same root is also found in the words physics, physician, and physical. Although many think physics refers to how energy and matter work, it may also refer to how nature functions. Physiology can be further divided into subfields such as animals, bacteria, plants, etc. But in this study guide, we will focus on how human beings function.

Organisms can be further broken down into various organization levels, all studied by physiologists. Each organism consists of several organ systems which work together to keep the organism alive. The organ systems within the organisms usually consist of several glands and organs. Each organ is a unique structure with a particular function. Organs consist of tissues. Tissues consist of cells that are similar in structure and function. As there are a lot of terms to remember in anatomy, you will also have to memorize several physiology terms. But before that, let's review the definition of physiology.

What is Physiology?

Physiology refers to the study of normal functioning within living organisms. This study is one of the subsections of biology, which covers a wide range of topics, including biological compounds, anatomy, cells, organs, and how all of these interact with one another to maintain the life of the organisms. From the most ancient theories to the modern techniques in the molecular laboratory, research into physiology has significantly shaped how we understand the parts of our body, how these parts communicate with each other, and how they all work together to keep us alive. Here are some critical points about this subject to remember:

- Physiology is sometimes considered a study of all the processes and functions responsible for creating life.
- The study of this subject dates to 420 BC.
- The study of this subject can be divided into several disciplines which cover varying topics like defense, evolution, exercise, and so on.

In another sense, physiology covers a wide range of disciplines and topics within the biology of human beings and beyond. For yet another reason, studying this subject is also considered the study of life. In it, you would ask many questions about how organisms' internal structures work and their interactions with the outside world. Physiology also tests how the systems and organs within the bodies of organisms function. And how they communicate, and combined efforts make it possible for the organisms to survive.

PHYSIOLOGY TERMS

Specifically, human physiology can be subcategories covering a vast amount of important information. Physiology researchers may focus on studying anything from specific things like microscopic organelles to broader topics like ecophysiology. Human physiology examines biological systems at the cellular level, organism level, and everything else in between. The primary systems of the human body which are focused on in physiology are:

- **Circulatory system** which includes the blood vessels, heart, blood properties, and the process of circulation for healthy and sick individuals.

- **Digestive and excretory system** which include the movement of food from the mouth all the way to the anus along with studying the liver, pancreas, spleen, and how food is converted into energy and how it finally exits from the body.

- **Endocrine system** which includes studying endocrine hormones that transmit signals throughout the body along with the major endocrine glands which are the thyroid, pituitary, gonads, parathyroids, and adrenals.

- **Immune system** which includes the natural defense system of the body consisting of the thymus, lymph system, and white blood cells. A complex assortment of molecules and receptors combine for the purpose of combating pathogens to keep the body protected.

- **Integumentary system** which includes the nails, hair, skin, sebaceous glands, and sweat glands.

- **Musculoskeletal system** which includes the whole skeleton along with all the muscles, cartilage, tendons, and ligaments. The bone marrow is also part of this system where red blood cells are produced, and how phosphate and calcium are stored in the bones.

- **Nervous system** which includes the spinal cord and brain (central nervous system) along with the peripheral nervous system. It also includes the study of the senses, thoughts, movements, emotions, and memories.

- **Renal or urinary system** which includes the ureters, kidneys, urethra, and bladder. This is the system which creates urine, eliminates water from blood, and carries waste away.

- **Reproductive system** which includes the sex organs and the gonads. Studying this system also includes learning about how a fetus forms and how the fetus is nurtured in the womb for the whole gestation period.

- **Respiratory system** which includes the trachea, lungs, nasopharynx, and nose. This is the system which brings oxygen into the body while expelling water and carbon dioxide.

The focus of Human physiology is studying the body's systems. However, physiology also includes several disciplines. Some examples of these disciplines are:

Cell physiology

Cell physiology refers to the study of how cells function and interact. This discipline mainly focuses on neuron transmission and membrane transport.

Systems physiology

Systems physiology is the mathematical and computational modeling of highly complex biological systems. This discipline describes how individual cells come together to respond as an entire system. Here, you will learn more about cell signaling and metabolic networks.

Evolutionary physiology

Evolutionary physiology refers to the study of how systems or parts of these systems have changed and adapted over time. Research on this discipline is very broad as it includes the role of behavior in sexual selection, physiological changes, and evolution in terms of geographic variation.

Defense physiology

Defense physiology refers to the changes that occur as a reaction to threats, like when the body prepares for a "fight or flight" response.

Exercise physiology

Exercise physiology refers to the study of physical exercises. The topics include research into:

- Biochemistry
- Bioenergetics
- Biomechanics
- Cardiopulmonary Function
- Skeletal Muscle Physiology
- Hematology
- Nervous System Function
- Neuroendocrine Function

These are just some of the disciplines which are part of physiology. There are several others. No matter what you plan to focus on while studying physiology, there are certain terms you should remember if you want to understand the topics better. Let's go through these terms now.

PHYSIOLOGY TERMS

Now that you have reviewed the definition of physiology and then some, it's time to go through some of the most important terms used in this subject. So here, we will go through the most used terms in physiology to help you understand and remember them better.

Abduction refers to the movement away from a median plane. Legs are considered in **abduction** for a person who stands with feet apart.

Abduction describes the movement wherein a part of the body moves away from the middle line. Your fingers are in **abduction** when you spread them out.

Abduction describes a motion going farther from the median plane of the body. Raising your arm places, it in **abduction**.

Active movement refers to movements done actively without help. Therapy can help patients move from bed rest to **active movement**.

Active movement describes a movement which is actively performed without assistance. Healthy people can do **active movement** without difficulty.

Active movement refers to movements a person does actively without any help or assistance. People who age may find **active movement** more challenging.

Adduction refers to the movement toward a median plane. Squeezing your knees together causes **adduction**.

Adduction describes the movement wherein a part of the body moves toward the middle line. Your fingers are in **adduction** when you press them against each other.

Adduction describes a motion going toward the median plane of the body. Pressing both arms against the sides of your body places them in **adduction**.

Anatomical planes refer to the imaginary planes that pass through the human body. The **anatomical planes** include the horizontal, median, coronal, and sagittal planes.

Anatomical planes refer to the non-existent planes passing through the body while in an anatomical position. Physiologists use the **anatomical planes** when describing movements of the body.

Anatomical planes refer to the imaginary planes used to describe body positions and movements. The **anatomical planes** are also used frequently in other medical subjects.

Anatomical position refers to the position of the body used as the main reference when describing body parts. Medical professionals use the **anatomical position** no matter what position their patient is in.

Anatomical position refers to the body's position used as the main reference when describing body parts and how they relate with each other. The **anatomical position** helps reduce confusion when describing injuries.

Anatomical position refers to the position of the body used as a standard method for documenting the position of body parts in relation to each other. In the **anatomical position**, the palms are facing forward.

Arthritis is a type of inflammatory condition affecting the joints. Obese individuals are more susceptible to **arthritis**.

Arthritis refers to an inflammatory condition that can be autoimmune, traumatic, or infectious in origin. A person who suffers from **arthritis** requires a specialized workout.

Arthritis refers to a painful condition which causes the joints to swell. There are topical pain relievers which may help ease the pain caused by **arthritis**.

Bone refers to the hard and calcified tissue which make up the skeleton. The older people get, the more brittle their **bones** become.

Bone refers to a calcified tissue consisting mainly of gelatine, calcic carbonate, and calcic phosphate. The skull is a **bone,** too.

Bone is the hard and calcified tissue made up largely of gelatine, calcic carbonate, and calcic phosphate. Each **bone** in our skeleton serves a purpose.

Bursitis refers to a condition wherein a bursa gets inflamed. Most cases of **bursitis** occur in the subdeltoid bursa.

Bursitis refers to the inflammation or swelling of a bursa. Traumatic olecranon **bursitis** commonly occurs after a contusion.

Bursitis describes a condition wherein a bursa swells. Certain exercises aren't recommended for those who suffer from active shoulder **bursitis**.

Carpal tunnel syndrome refers to a condition that occurs when a disturbance happens in the function of the wrist's median nerve.

Some pregnant women are more susceptible to **carpal tunnel syndrome**.

Carpal tunnel syndrome is a condition where the wrist's median nerve function gets disrupted as it passes through the carpal tunnel. **Carpal tunnel syndrome** may be alleviated through relaxation and breathing methods.

Carpal tunnel syndrome is a condition that affects the wrist as the function of its median nerve gets disrupted. In severe cases of **carpal tunnel syndrome**, the pain may radiate to the hands and arms.

Carpal tunnel release refers to a surgical procedure performed to relieve carpal tunnel syndrome. A person may opt to have **carpal tunnel release** when the pain caused by carpal tunnel syndrome becomes unbearable.

Carpal tunnel release is a type of orthopedic procedure which involves surgery to relieve the pressure on the wrist's median nerve. **Carpal tunnel release** is the best option for those who suffer from severe carpal tunnel syndrome.

Carpal tunnel release refers to an orthopedic procedure done to relieve the pressure which is exerted on the wrist's median nerve in the carpal tunnel. The incision for a **carpal tunnel release** is done at the base of the patient's hand.

Cervical spine refers to the vertebrae found in the neck bones. Kyphosis is a type of progressive disorder that affects the **cervical spine**.

Cervical spine is the seven vertebrae which serve as the framework for the neck bones. It's possible to immobilize the **cervical spine** using tape, a sandbag, and a cervical collar.

Cervical spine is seven vertebrae in the neck which make up the bones. Cervical spondylosis is a degenerative condition which affects the **cervical spine**.

Circumduction is a circular motion which combines extension, flexion, adduction, and abduction. The normal range of motion for the hips includes **circumduction**.

Circumduction is a type of circular movement combining extension, flexion, adduction, and abduction. A patient displays **circumduction** when they swing out their legs while walking.

Circumduction is a circular movement which combines extension, flexion, adduction, and abduction. **Circumduction** occurs when you move your arms around in a circular motion.

Contracture is a type of condition of high resistance that's fixed to a muscle's passive stretch. **Contracture** may cause joint deformity.

Contracture refers to a condition of fixed high resistance when a muscle is passively stretched that's caused by tissue fibrosis. When scar tissues form around breast implants then contracts, this may cause capsular **contracture**.

Contracture is a condition of high resistance that's fixed when a muscle is passively stretched that's caused by tissue fibrosis. Physiotherapy may help prevent joint **contracture**.

Contusion is bleeding that happens within muscles. On the surface of the skin, a **contusion** may appear as a bruise.

Contusion occurs when there is bleeding in a muscle because of a direct blow. One of the most common sports injuries is a **contusion**.

Contusion happens when a direct blow to one of the muscles causes bleeding. Some MRI scans may show a **contusion** along with other muscle damage.

Cryotherapy is the therapeutic utilization of cold to help limit the progression of edema and to reduce discomfort. **Cryotherapy** is a type of counterirritation.

Cryotherapy is the process of using cold as part of treatment to reduce discomfort. The success of **cryotherapy** may vary depending on the accompanying treatment regimen.

Cryotherapy involves using cold therapeutically to limit the progression of edema. **Cryotherapy** may have some harmful side effects.

Depression refers to a reduction or lowering of a specific function or biological variable. In physiology, **depression** is the opposite of elevation.

Depression occurs when an organ function or a biological variable is lowered. A **depression** of the thyroid gland may cause some deleterious effects.

Depression is the lowering or reduction of a specific organ function or biological variable. **Depression** of our central nervous system may cause drowsiness or unconsciousness.

Elevation refers to an upward movement. **Elevation** happens when you shrug your shoulders.

Elevation is an upward movement. **Elevation** of the feet can help reduce swelling.

Elevation happens when you move a part of your body upward. Avoid excessive **elevation** when nursing a bitten limb.

Embolus refers to a clot that forms when leukocytes or platelets block a blood vessel. The bloodstream carries an **embolus** until it gets lodged somewhere causing a blockage.

Embolus is a clot formed by leukocytes or platelets blocking a blood vessel. When the blockage happens in the lungs, it's called a pulmonary **embolus**.

Embolus refers to a clot that is formed by leukocytes or platelets blocking a blood vessel. A blood clot is the most common type of **embolus**.

Extension refers to the straightening of a joint to make it longer or larger. The knees undergo **extension** when the legs are straightened.

Extension happens when a person straightens the joint to make it larger or longer. **Extension** of the ankle happens during stretching exercises.

Extension is the straightening of a joint to enlarge the ankle. The elbows undergo **extension** when you stretch your arms.

Flexion refers to the bending of a joint which makes the angle between bones smaller. **Flexion** of your fingers happens when you close your fist.

Flexion occurs when bending a joint minimizes the angle between two separate bones. **Flexion** occurs when you bend your knees.

Flexion is when you bend a joint to make the angle between two separate bones smaller. **Flexion** happens when you bend your arm.

Fracture is a rupture or break in a bone's cortex. A **fracture** must be treated immediately.

Fracture occurs when the cortex of a bone breaks or ruptures. For athletes, a **fracture** can be an extremely devastating injury.

Fracture refers to a rupture or break in the cortex of a bone. When a child gets a **fracture**, it can be particularly worrisome.

Hemarthrosis refers to bleeding which originates from inside a joint. A clotting screen may be needed to diagnose **hemarthrosis**.

Hemarthrosis is bleeding which originates from inside an injured or damaged joint. **Hemarthrosis** isn't easily detectable from the outside.

Hemarthrosis occurs when bleeding happens originating from inside a joint which has been injured or damaged. **Hemarthrosis**

may cause a chemical reaction to happen between the blood and the cartilage and synovium.

Inflammation is a swelling caused by illness or injury. There are many types of food which may cause **inflammation**.

Inflammation refers to a localized swelling resulting from a protective response by the body. Chronic **inflammation** can cause a lot of pain.

Inflammation refers to a swelling caused by an injury or illness. There are also different kinds of food which help reduce **inflammation**.

Irritability is the behavior of a specific condition when provoked. **Irritability** may occur when something in the environment changes.

Irritability is a type of behavior caused by provocation. Aside from being a physiological reaction, **irritability** can also mean excessive sensitivity.

Irritability refers to the behavior of a condition when provoked. Often, **irritability** is expressed as frustration or anger.

Ligament is a band of tissue which connects cartilages or bones. Damages to the **ligaments** are very common.

Ligament refers to a band of fibrous tissues connecting cartilages or bones. Radiography may be used to diagnose **ligament** deficiencies.

Ligament is a band of tissue which connects cartilages or bones which serves to strengthen and support the joints. Surgery may be needed to reconstruct a damaged **ligament**.

Lumbar spine refers to the vertebrae which serve as the frame of the lower back. Stress fractures are common in the **lumbar spine**.

Lumbar spine is made up of 5 vertebrae in the lower back. Measuring the **lumbar spine** and femoral necks can help determine density mass.

Lumbar spine consists of 5 vertebrae which make up the bones of the lower part of the back. The sacrum is located below the **lumbar spine**.

Nerve refers to an elastic bundle of fiber with accompanying tissue. A **nerve** transmits nervous impulses between the nerve centers and the different parts of the body.

Nerve is an elastic bundle of tissue and fibers with a whitish color. Each **nerve** in the human body is important.

Nerve is an elastic, whitish bundle of fiber with accompanying tissue. Hair doesn't contain a single **nerve**.

Opposition refers to a hand movement where the thumb and fifth finger touch. Not all animals can perform **opposition**.

Opposition is a hand movement where the thumb and fifth finger meet. Only primates can do **opposition**.

Opposition refers to the movement of the hand wherein the thumb would touch the fifth finger. **Opposition** is a unique hand movement.

Osteitis refers to a condition wherein the bones start to soften. **Osteitis** occurs because of impaired mineralization.

Osteitis is a condition marked by bone softening. **Osteitis** typically comes with tenderness, pain, muscular weakness, and other symptoms.

Osteitis occurs when the bones start softening. **Osteitis** may also occur in people who are deficient in calcium and vitamin D.

Osteomalacia refers to bone inflammation because of a pyogenic organism. **Osteomalacia** may be localized, or it may spread to the other parts of the bone.

Osteomalacia is a condition where bones get inflamed because of a pyogenic organism. **Osteomalacia** may occur when there is a sudden drop in the calcium levels in the blood.

Osteomalacia is the inflammation of a bone because of a pyogenic organism. **Osteomalacia** may also increase a person's susceptibility to fractures.

Osteoporosis refers to the decrease in the bone mass amount. Older individuals are more prone to **osteoporosis**.

Osteoporosis occurs when the amount of bone mass is reduced. If you want to prevent **osteoporosis**, make sure to get adequate amounts of calcium and vitamin D.

Osteoporosis refers to the decrease in the bone mass amount which may lead to a higher susceptibility to fractures. Heavy metals might make **osteoporosis** more severe.

Passive mobilization is a type of treatment technique done manually to increase movement or reduce pain in one of the joints. There are different grades of **passive mobilization**.

Passive mobilization is a manual treatment method to help reduce joint pain or increase joint movement. The grade of **passive mobilization** to be performed depends on the desired outcome.

Passive mobilization refers to a manual technique for treatment to increase movement or reduce pain in one of the joints. The grade of passive mobilization to be performed depends on the patient's condition.

Passive movement is a movement of one of the body parts which happens because of an external force. Passive movement may happen during therapy sessions.

Passive movement occurs when one of the body parts moves as a result of an external force. **Passive movement** is done by therapists without assistance from a patient.

Passive movement happens when an external force causes a movement of one part of the body. **Passive movement** is a type of involuntary movement.

Pronation refers to when the forearm moves with the palm facing anteriorly. **Pronation** occurs when you unscrew a screw using your right hand and forearm.

Pronation is a movement of the forearm wherein the palm is facing anteriorly. Some exercises require **pronation**.

Pronation is when the palm faces anteriorly when the forearm moves. There are certain injuries which make **pronation** impossible.

Protraction refers to an anterior movement. **Protraction** happens when you stick out your chin.

Protraction means moving anteriorly. The scapula involves **protraction** and retraction along the thoracic wall.

Protraction occurs when you move a body part anteriorly. The tongue's hydrostatic elongation requires **protraction** and lengthening.

Range of movement is how much movement is made at a joint expressed in degrees. Exercise is important to maintain a good **range of movement.**

Range of movement refers to how much movement is made at a joint. As a person grows older, his **range of movement** may decrease.

Range of movement is the amount of movement made at a joint expressed in degrees. Injuries may cause a limited **range of movement**.

Referred pain is a pain that's felt far from the actual pain source. A toothache may cause **referred pain**.

Referred pain is a type of pain a person feels away from the real source of pain. **Referred pain** may be felt when it radiates from an injury.

Referred pain occurs when a person feels pain far from the actual pain source. Some types of lower back pain may also cause **referred pain** in the legs, buttocks or groin.

Retraction refers to a posterior movement. Squeezing your shoulder blades towards each other causes a **retraction** of the shoulders.

Retraction means moving posteriorly. **Retraction** of the shoulders occurs in various gymnastic exercises.

Retraction occurs when you move a body part posteriorly. The only parts of the body capable of retraction are the jaw and shoulders.

Rickets is a condition suffered by those who have a vitamin D deficiency. **Rickets** is more common in young children than in adults.

Rickets commonly occurs in infants and children who are deficient in vitamin D. The amount of vitamin D in breast milk isn't enough to prevent **rickets**.

Rickets refers to a condition caused by a vitamin D deficiency. **Rickets** often results in a child becoming bow-legged.

Rigidity refers to abnormal inflexibility. **Rigidity** occurs after death.

Rigidity refers to abnormal stiffness. **Rigidity** may be a symptom of a more serious condition.

Rigidity is abnormal inflexibility or stiffness. Not getting enough exercise may lead to **rigidity**.

PHYSIOLOGY TERMS

Sciatica is a type of syndrome which causes radiating pain. **Sciatica** is the pain that's felt along the sciatic nerve.

Sciatica is a condition classified as a syndrome which is characterized by radiating pain. A herniated disc in the lumbar region may cause **sciatica**.

Sciatica is a syndrome which is characterized by radiating pain. Severe **sciatica** requires immediate treatment.

Spasticity refers to a hypertonicity state. There are new drugs which may help treat **spasticity** in children.

Spasticity occurs when there is an increase of the normal muscle tone (hypertonicity) with a heightening of deep tendon reflexes. One of the symptoms of tetanus is **spasticity**.

Spasticity refers to the state of hypertonicity or an increase in the normal muscle tone. Severe **spasticity** can be disabling.

Sprain is a type of joint injury. Playing sports may cause a **sprain**.

Sprain is an injury which affects the joints, specifically the structures within the joints. A **sprain** isn't as severe as a fracture.

Sprain is a joint injury wherein stabilizing structures such as ligaments are torn or damaged. A **sprain** typically comes with swelling and pain.

Strain is a type of muscle injury. The most common causes of **strain** are over contraction or overstretching.

Strain is an injury which affects the muscles, specifically the fibers within the muscles. Be very careful when exercising so as not to experience **strain**.

Strain is a muscle injury wherein the fibers are torn or damaged. Severe **strain** comes with high levels of pain.

Subluxation is an abnormal bone movement which ends up compromising a joint. **Subluxation** is also known as a partial dislocation.

Subluxation refers to an abnormal movement of a bone which compromises the adjacent joint. Severe arthritis may cause **subluxation.**

Subluxation refers to an abnormal bone movement which compromises the adjacent joint. **Subluxation** occurs when one of the vertebrae gets dislocated.

Supination refers to when the forearm moves with the palm facing posteriorly. You may experience pain when you perform **supination** under resistance.

Supination is a movement of the forearm wherein the palm is facing posteriorly. Some exercises require **supination**.

Supination is when the palm faces posteriorly when the forearm moves. There are certain injuries which make **supination** impossible.

Tendinitis refers to an inflammation of the tendons and their muscle attachments. **Tendinitis** may cause a limitation of movement of the affected area.

Tendinitis is a condition wherein the tendons and their muscle attachments get inflamed. **Tendinitis** is a condition which causes pain depending on its severity.

Tendinitis occurs when the tendons and tendon muscle attachments get swollen or inflamed. **Tendinitis** must be treated to avoid the rupture of the affected tendon.

Tendon is a strong and fibrous connective tissue. When a **tendon** gets torn, the person won't be able to move his joint.

Tendon refers to a strong and fibrous connective tissue connecting bones and muscles. A torn **tendon** takes a long time to heal.

Tendon is a fibrous and strong connective tissue connecting muscles to bones. Aerobic exercises may help improve the circulation of blood to the bursa and **tendon**.

Thoracic spine refers to the vertebrae which serve as the frame of the region of the upper and mid-back. Each vertebra in the **thoracic spine** is attached to the ribs.

Thoracic spine is made up of 12 vertebrae in the region of the upper and mid-back. Vertebral bodies with an ovoid shape can be found in the **thoracic spine**.

Thoracic spine consists of the 12 vertebrae which make up the bones of the region of the upper and mid-back. The **thoracic spine** is part of the thoracic cage.

Ultrasound is a type of electrotherapy technique used for the treatment of various conditions. **Ultrasound** makes use of sound waves with a high frequency that pass through the skin.

Ultrasound refers to an electrotherapy treatment method typically used by physiotherapists. **Ultrasound** may help reduce inflammation, break scar tissue down, and stimulate healing.

Ultrasound is a common treatment technique which uses electrotherapy. There is a special device used for **ultrasound**.

CHAPTER 3

Physiology Terms glossary

Physiology is a big, complex, and heavy topic and comes with its own extensive, complex glossary of terms. But, don't stress too much on all these terms. You can always refer to this book. In fact, the more you do, the more these terms will stick in your mind. So, take your time.

Oh, and a quick side note: We will cover stand-alone terms here, as there will be a separate chapter focusing on **suffixes and prefixes.**

Ok let's begin:

A

Ablation - the surgical removal of a body tissue. Catheter **ablation** is a surgical procedure in which a small area of heart tissue that causes irregular heartbeats is removed with radiofrequency energy.

A band - the region of a sarcomere (basic unit of the striated muscle tissue) that contains thick filaments of myosin. It represents the entire length of the thick filament.

Abscissa - is the X-axis or horizontal axis in a graph, along which are plotted the units of variables used in the study, for example in a time-temperature study.

Absolute refractory period - a period during the action potential during which no stimulus, no matter how big, can trigger another action potential from an excitable tissue. Ex: The action potential reaches its peak while in the **absolute refractory period**.

Absorption - the transfer of substances into the body or into the blood. Ex: **Absorption** through the skin is the transport of chemicals from the outer layers of the skin into the deeper ones and into circulation.

Acclimatization - the process of adapting to a new environment, for example with altered pressure or temperature. It's usually used to show adaptation to a single changed factor.

Accommodation - a process of adapting to new conditions, and it can relate to getting used to a stimulus until you become unresponsive to it (such as a smell) or to the adjustment of the eye focus.

Accumulation - a progressive build-up of substance. Ex: The **accumulation** of toxins in the organism is more dangerous than being exposed for a short period of time to toxic substances.

Acetylcholine (abbreviated **ACh**) - a chemical neurotransmitter used by both the central and the peripheral nervous system. Ex: **Acetylcholine** exerts its effects by activating receptors located on the surfaces of the cells.

Acetylcholinesterase - an enzyme that degrades acetylcholine to choline and acetate, terminating its action at the synapse.

Acid - a solution in which the solvent is water (aqueous solution), with a pH that's less than 7.0

Acidosis - an abnormally low pH of the blood

Acid-Base balance - the control of the factors that influence the pH of the blood. This balance can be influenced by altered respiration, metabolism, and absorption or by disturbances in the function of the system (for example becoming ill).

Action Potential (a.k.a electric impulse or nervous impulse) - a change of voltage recorded inside or close to a nerve or other excitable cell, due to a characteristic change in membrane potential. Action potentials come as a reaction to a stimulation.

Activation - the induction of action potentials in the cell membrane

Activation energy - the energy needed to trigger an action potential

Active - something that requires energy (such as the active transport) or a nerve fiber that is undergoing action potentials.

Active site - the part of a cell or a receptor that binds to a substrate. Ex: The **acetylcholine** binds to the active site of the receptors.

Active Transport - the movement of a substance across a membrane, with the help of a carrier or by using energy from a process such as hydrolysis or the movement of another substance.

In other words, it's a form of transport that can't happen on its own.

Acuity - the ability to identify the finer details in a sensory pattern. Ex: We test our visual **acuity** with the help of the Snellen Chart (or the eye chart).

Acute - something sudden, sharp, and severe. It can be an **acute** pain, an **acute** disease, or an **acute** response to stimulation.

Adaptation - the decline in a response while a stimulus is constantly maintained. For example, our eyes adapt to different light levels.

Adequate stimulus - a form of stimulus that is big/strong enough to trigger a response such as an action potential or a reflex.

Adrenaline (or **epinephrine**) - a hormone secreted by the adrenal gland in conditions of stress. It acts as a neurotransmitter. It's also used as medication (**epinephrine**), especially in cases of cardiac arrest. Ex: Extreme sports trigger an **adrenaline** rush.

Adrenergic - related to adrenaline (or noradrenaline)

Adenohypophysis - the anterior region of the pituitary gland

Adrenarche - an early sexual maturation stage that precedes puberty

Aerobic - a process that requires the presence of oxygen, for example, aerobic metabolism

Aetiology (or **etiology**) - the cause or "source" of a disease. Ex: Malaria's **aetiology** is a protozoan that enters the human body through the bite of an Anopheles mosquito.

Afferent - traveling or transmitting something towards a region or structure of interest. Ex: **Afferent** nerves travel towards the central nervous system.

Affinity - the strength of the link between two chemicals; a chemical and an ion or a receptor and an enzyme

Afterdischarge - action potentials that occur after the stimulus stopped

Agonist - something that mimics or assists an action. For example, a molecule that binds to a receptor and activates a response like that of a natural ligand is called an **agonist**.

All-or-None/Nothing - this term is used when describing the action potential, and it means that an action potential only occurs if the stimulus exceeds a threshold level.

Amenorrhea - the absence of a menstrual cycle in a sexually mature female individual

Amiloride - is a diuretic (water pill) that prevents the body from absorbing too much salt, and maintains normal potassium levels

Anesthesia - the absence of sensation. **Anesthesia** can be general (the subject is unconscious) or local (lack of sensation only in a specific part of the body).

Analgesia - the absence of pain sensation

Analogue - in medicine, an analog drug or hormone is one that has a similar action/structure/composition to another

Aneurysm - an abnormal bulging part of a vessel (usually of an artery) caused by a weak spot in the blood vessel wall

Anion - a negatively charged ion

Anode - a positive electrode

Anoxemia - the lack of oxygen or insufficient oxygen in arterial blood

Anoxia - a condition without oxygen; the absence or deficit of oxygen

Antagonist - something that has the opposite effect on something else. For example, a muscle that pulls in the opposite direction to another; a chemical that has an opposite action to another.

Anterograde - moving along an axon, towards the axonal terminal and away from the neuronal cell body

Antiport - exchange; a membrane transport protein that transports two different molecules, in the opposite direction, at the same time.

Anucleate - without a nucleus. Ex: Red blood cells are **anucleate**.

Aqueous - something that is associated with water (*aqua* is the Latin for water)

Arteriosclerosis - the loss of elasticity of arteries, due to aging or disease

Arteriovenous - relating to arteries and veins

Asynchronous - out of sync, not having an ongoing rhythm.

Atropine - an inhibitor of **acetylcholine** receptors

Autonomic - related to the autonomic nervous system that innervates smooth muscles, visceral organs, and glands, which are not under voluntary control. it also relates to **autonomic** reflexes, which control actions such as heart rate, sweating, peristalsis, and urination.

Autoregulation - the ability of an organ or tissue to regulate its own functions without the need for external input. For example, the **autoregulation** of the kidneys ensures a constant blood flow and pressure to keep a constant glomerular filtration rate (GFR).

Axon - a nerve fiber. Ex: A nerve is made up of multiple **axons**.

Axon hillock - the origin of the axon, the part of the neuron cell body from which the axon emerges. Ex: The **axon hillock** is the site of action potential initiation.

B

Ball and Chain Model - a model created to explain the inactivation of some voltage-gated ion channels. Both the "ball" and the "chain" are formed by amino acids. According to this model, the channel protein has a globular portion (the ball) attached to it (through the link), which "plugs" the open channel.

Baseline - a period of recording before a stimulus is given or before something happens. For example, it's impossible to get a **baseline** measurement on a patient before they get sick, so it's mostly done in scientific environments.

Biconcave - concave on both sides of the structure, mostly referring to lenses or discs. Also, mature red cells assume a **biconcave** shape to ensure that the blood flows properly through the vessels.

Bicuspid valve (a.k.a **left atrioventricular valve** or **mitral valve**) - a one-way valve, located between the left atrium and the left ventricle of the heart.

Block - in physiology, it means to stop something from happening or to diminish the action of a drug. For example, a local anesthetic **blocks** action potentials in nerve fibers.

Body Mass Index (abbreviated **BMI**) - this index is measured by dividing body weight (in kg) by body height (in meters) squared. it's mostly correlated with body composition, a higher **BMI** value meaning a higher body fat content. It's used to classify whether someone is underweight, normal, overweight, or obese.

Bulk flow - a flux of a solution that carries all its solvents, the solution as a whole.

C

C cell - another name for the parafollicular cell type found within the thyroid gland. **C cells** produce the hormone **calcitonin**.

Calorie - a unit of energy (heat), one calorie is equal to 4,18 Joules. it's not the same with the "calorie" used in nutrition, but to avoid confusion it is better to use Joules when measuring heat.

Cannula - a tube that's inserted into a blood vessel, so that solutions can be put into it or withdrawn from it. Ex: A **cannula** is used when transfusing blood.

Capacitance - in physiology, **capacitance** vessels are those that can contain an extra volume of blood without a major increase in pressure, mainly the veins.

Cardiac - related to the heart, for example, **cardiac** skeleton (thick tissue that separates the atria and the ventricles). The **cardiac** sphincter, however, is at the top end of the stomach and has nothing to do with the heart.

Cardiac output - the volume of blood pumped every minute by each ventricle of the heart. It's measured in liters per minute.

Carrier - in membrane transport, a **carrier** is a protein that binds a substance and transports it to the other side of the membrane by changing its conformation. **Carriers** are involved in active transport and facilitated diffusion. Some **carriers** can move two substances at once.

Catabolism - metabolism that involves the destruction or disappearance of tissue. For example, in cases of starvation, fat deposits are **catabolized**.

Catheter - a tube inserted into a narrow opening to remove (urinary **catheter**) or introduce fluid. A **catheter** can be introduced through a **cannula**.

Cation - a positively charged ion

Cerebral - relating to the brain

Cerumen - the scientific term for earwax

Cervical - relating to the neck or to the lowest part of the uterus that faces the vaginal canal (**cervix**)

Chalone - an endogenous molecule specific to the tissue, that inhibits mitosis (cell division) in the tissue that released it.

Channel - a membrane protein that has pores, through which ions, water, and polar molecules can pass to the other side of the membrane. **Channels** are mostly responsible for the permeability of membranes, but they are also very selective, each **channel** corresponding to a certain molecule or ion.

Channel inactivation - it refers to a change in the conformation of the channel protein, by which the channel goes from "open" to "closed". In other words, the ion, molecule, or substance that is generally allowed to pass through that channel, is now blocked from entering.

Chemical gradient - the concentration gradient of an ion or a molecule. This **chemical gradient** exists across biological membranes or in a solution that does not have a barrier separating the area of higher concentration from that of lower concentration. Free movement of substances/ions/molecules from an area of higher concentration to that of lower concentration.

Cholinergic - it refers to receptors, synapses, or neurons where the neurotransmitter used is **acetylcholine**

Chronic - something continuous, or that exists for a long time such as a **chronic** pain or disease. It's the opposite of **acute**.

Chyme - the partially digested contents of the stomach that pass into the small intestine (the duodenum to be more precise). It's mostly in a semi-solid state.

Circadian rhythm - a cyclic pattern of behavior or of tissue activity that repeats within a period of 24 hours. For example, some hormones (melatonin, cortisol prolactin, thyroid stimulating hormone), secret in a **circadian rhythm.**

Cofactor - a chemical that binds to an enzyme, and that is needed in order for a reaction to take place. A **cofactor** is essential for the activity of the enzyme.

Colloid - it's a protein-rich fluid, usually referring to the thyroid **colloid** that is present within the lumen of the thyroid follicles.

Compartment - a specific part of the total volume of a tissue or organism, for example, the intracellular fluid **compartment**.

Competitive - it means that two or more substances that are related or similar enough can bind to the same site of a receptor, enzyme, or carrier. If two such substances are present, the binding of both is reduced, leading to **competitive** inhibition (none can bind to its full potential because the other is present).

Compliance - it measures how easily something complies with a force that's been applied to it. For example, muscles and tendons will comply with a force by increasing in length.

Concentration - the quantity of a solute per volume unit. It is often expressed in moles per liter, which is why it's also called **molar concentration** or **molarity**.

Constriction - the narrowing of a vessel, for example, due to the **contraction** of a smooth muscle

Contraction - a state of activity of a muscle, in which it can produce force or shorten, depending on the mechanism that triggers the **contraction**.

Cortex - the outer part of an organ or the skin. For example, the cerebral **cortex** is the protective layer of the neural tissue.

Corticotroph - an endocrine cell of the anterior pituitary gland that synthesizes and releases **adrenocorticotropic hormone** (ACTH)

Cotransport - a secondary active transport in which a carrier protein couples two ions or an ion and a molecule and transports them against a concentration or electrochemical gradient.

Cranial - relating to the skull (**cranium**)

CT scan - a computer tomography (an axial tomography to be more precise). It shows the internal structure of tissues.

Curare - a blocker of nicotinic cholinergic receptors, found at neuromuscular junctures. Even in small doses, it can lead to muscle weakness.

Current -the flow of charge, carried by ions in solutions

Current clamp - an electrophysiological technique in which the current passing through the membrane is controlled and the voltage measured.

D

Deiodinase - an enzyme that converts thyroxine (T4) to triiodothyronine (T3, or the active form of the thyroid hormone). **Deiodinase** is found in target tissues of the thyroid hormones.

Dendrite - branching structures (that resemble tree branches) which are part of the neuron (the part that receives synaptic contacts).

Depolarisation - a change in the value of the membrane potential, either towards being more negative or positive than the resting potential.

Depression - a decrease in the activity of a tissue or in the amplitude of a response. If it happens due to the involvement of structures that have this effect, then it's called "**inhibition**". Not to be confused with "clinical depression", which is a mental condition.

Diastole - the part of the cardiac cycle in which the ventricular myocardium relaxes (so, immediately after **contraction**).

Diffusion - the process in which molecules get from one place to another, usually through a membrane. If a carrier is involved to aid the transportation of molecules, then it's called "**facilitated diffusion**".

Dilatation (or **dilation**) - the opening of a tubular or hollow structure. For example, a blood vessel or the eye pupil.

Dissect - to remove a piece of tissue or a cell from a larger structure, or to cut in order to expose the deeper structures.

Drug - a chemical that affects biologicals structures, usually referring to chemicals that are used clinically.

Dyspareunia - painful or otherwise difficult intercourse.

E

Edema - an accumulation of fluid in the interstitial compartment. It leads to visible swelling of the affected tissue.

Efferent - leading away from a region. For example, **efferent** fibers (neurons) transmit information from the central nervous system to the effector organ.

Efflux - the movement of a substance out of a cell. It's usually reported as a rate, for example, the amount of fluid that moves through the area in a specific amount of time.

Electrical gradient - the electrical potential that causes an ion to move in one direction or another in a solution.

Electrocardiogram (or EKG or ECG) - a recording of the electrical activity of the heart. More specifically, it shows the electrical events of the cardiac cycle, with the usage of electrodes.

Electrochemical gradient - the balance of the chemical and **electrical gradients** that act on an ion, moving it across the membranes.

Electroencephalogram - it records voltage changes created by brain activity with the usage of electrodes.

Electrolyte - a salt that dissociates into ions, when you dissolve it in water. It can be strong or weak, and all tissue fluids have **electrolytes**.

Electromyogram - it records voltage changes in a muscle, with the usage of electrodes.

Electrogenic - an electrogenic process leads to the translocation of net charge across the membrane

Embarrass - it means preventing an organ or tissue from functioning normally. We call it feeling numb, for example, when we sit on a leg for far too long.

Embolism - the blockage of an artery by an embolus (abnormal fragments of material, such as thrombocytes, air bubbles, or fat). Venous **embolism** is usually called pulmonary **embolism**, meaning that the blockage happens at the level of the pulmonary circulation.

Endocrine cell - cells that produce or release hormones into the bloodstream. A group of **endocrine cells** is called an **endocrine gland**.

Enzyme - a protein that enhances the rate of a chemical reaction, be it to make it faster or more powerful.

Equilibrium - it refers to a condition in which there is no tendency for a substance, ion, or molecule to spontaneously move into a certain direction or another.

Equilibrium potential - the value at which a certain ion would be in a state of **equilibrium**. Each ion has its own **equilibrium potential**.

Euthyroidism - the normal state of function for the thyroid gland.

Excise - to remove or to cut, not necessarily to get to a deeper structure.

Exchange (or **antiport**) - it's a secondary active transport across a membrane, in which a protein couples two ions that move in opposite directions. The ion that moves down is known as the "driving" ion, while the other is the "driven" ion or molecule.

Excitable cell - the ability of some cells to be electrically stimulated and generate an action potential. Such as neurons, muscle cells, and some endocrine cells.

Excretion - the elimination of waste substances from the body. For example, the kidneys and the lungs are the two major organs that have an excretory function.

Exophthalmia (or exorbitism) - the protrusion or bulging of one or both eyes, caused by swelling, trauma, or disease.

F

Feedback - information that was sent to an earlier stage. **"negative feedback"** reduces the signal, while a "positive **feedback**" increases the signal.

Fibrillation - a repetitive high-frequency spontaneous contraction of a muscle, usually the cardiac muscle.

Flux - the rate of movement of a substance across an interface such as a membrane or an epithelial sheet.

G

GABA - another name for γ-Aminobutyric acid

Glucosuria - the presence of glucose in the urine, which is abnormal

Goiter - an enlargement of the thyroid gland due to iodine deficiency

Gonadotroph - the endocrine cell of the anterior pituitary gland, that synthesizes and releases follicle stimulating hormone (FSH) and luteinizing hormone (LH)

Graded potential - it refers to potentials that can vary in amplitude in direction, be it a synaptic or receptor potential

H

H zone - the region of a sarcomere, that is at the center of the A band and it consists only of thick filaments. The **H zone** becomes smaller as the sarcomere shortens during muscle contraction.

Halitosis - the scientific term for bad breath

Hirsute (or **Hirsutism**) - excessively amount of hair on the face and/or body

Hodkin cycle - a positive **feedback** loop in neurons, where an initial membrane depolarization leads to rapid membrane depolarization to achieve its **equilibrium** potential.

Hormone - a chemical messenger secreted by endocrine cells

Hyperacusis (or **Hyperacousis**) - an over-sensitivity towards moderately loud sounds

Hypercalcemia - a calcium level (in the plasma) that is above the normal value. **Hypocalcemia** is the opposite, with the level being under the normal value.

Hyperglycemia - a concentration of glucose in the plasma that is above the normal value. It's usually a sign of diabetes. **Hypoglycemia** is the opposite, with the concentration of glucose being under the normal value.

Hyperphagia - excessive consumption of food

Hyperplasia - it can refer to an increase in the number of cells of a tissue or to the enlargement of an organ or body part

Hyperthyroidism - a pathological condition in which the thyroid produces and releases an abnormally high level of hormones

Hypertrophy - an increase in the size of tissues or organs, due to the increase in the size of the cells

Hypovolemia - a decrease in the total blood volume, caused usually by blood loss

Hypoxemia - low oxygen level in the arterial blood. **Hypoxia** refers to low oxygen levels in the whole body (tissues) or in a general region (such as the brain).

I

I band - the region of a sarcomere that contains thin filaments. It's the lightest region of the sarcomere.

Impermeable - not allowing the passage of substances, referring to the membrane and its channels

Impermeant - the incapacity of a substance to pass through a biological membrane, despite the membrane not being **impermeable**.

Infarct - a dead or not functioning region of a tissue

Interstitial - the space in-between cells

Inward current - a negative current value or a downward deflection of a current

Iodine trap - the ability of a thyroid gland to accumulate iodine, against a high electrochemical gradient

K

Korotkoff sounds - this term refers to the sounds that can be heard when you place a stethoscope over a partially compressed peripheral artery. The sounds resemble "tapping" and are created by the pulsation of the blood through the semi-constricted artery

L

Lactotroph - an endocrine cell of the anterior pituitary gland that synthesizes and releases prolactin (PRL)

Lidocaine - a local anesthetic that is also used as an antiarrhythmic drug

Lyse - to dissolve or destroy, often referring to the "bursting" of cells

M

M line - the region of a sarcomere, on which the thick filaments are attached. It's located in the center of the sarcomere

Membrane potential - the voltage differences between the inside and the outside of a membrane. Generally, on the inside, it's negative, and positive on the outside

Metabolism - a series of chemical reactions that occur in the body, in relation to chemicals that are ingested or manufactured in the body

Motor unit - the "team" that encompasses the motor neuron and all the muscle fibers that it innervates

Myxedema - a condition characterized by thick skin, edema, and a decreased metabolic rate; that is associated with **hypothyroidism** (the thyroid does not secrete enough hormones).

N

Nascent - something that is recently synthesized or in the process of being synthesized

Nerve impulse (or **nervous impulse**) - is the same with action potential

Net flux - represents the amount of substance that was moved in or out of a cell

Neurotoxin- a chemical that has a nocive effect on neuron function, disrupting the normal function of the whole nervous system

Neurotransmitter - chemical messengers released by the neurons

O

Occlusion - a blockage. For example, of a blood vessel

Oliguria - an abnormally low production of urine (less than 400 mL per day)

Oophorectomy - the surgical removal of one or both ovaries

Outward current - a positive or upward deflection of a current

Overshot - the stage of the action potential in which the membrane potential is positive

P

P wave - a component on the **EKG** showing the depolarization of atrial myocytes

Parafollicular cell - cells found within the thyroid gland

Phonocardiogram (abbreviated **PCG**) - a recording of the sounds produced by the heart during the cardiac cycle

Plasma - the fluid portion of the blood that represents about 40 to 60% of the whole blood

Postprandial - the period after a meal

Presbycusis - the gradual loss of hearing, associated with aging

Puberty - a period of growth and change caused by hormonal changes, which leads to sexual maturity

Pulse pressure - this parameter is equal to the difference between systolic pressure (the highest blood pressure) and diastolic pressure (the lowest blood pressure)

Q

QRS complex - a component on the **EKG** showing the depolarization of ventricular myocytes

R

Reflex - a response to a stimulus, in which the signal is transmitted by neurons

Repolarization - the return of the membrane potential towards its normal value, after depolarization

Retrograde - the movement from the axon's extremities towards its neuronal cell body

S

Salpingectomy - the surgical removal of the fallopian tube

Sarcolemma -the plasmatic membrane of a muscle cell

Secretion - the cellular release of substances to the external environment of the cell

Spike phase - the stage of the action potential in which the membrane is rapidly undergoing depolarization

Synaptopathy - any pathological condition that leads to an abnormal function of the synapses

T

T wave - a component on the **EKG** showing the repolarization of the ventricles

Tachyphylaxis - a condition where a person develops a rapid tolerance to a drug, following a single or repeated administration

Tastant - chemical substances that provoke "taste" if they are dissolved in ingested liquids or saliva

Tinnitus - the condition in which a person hears a sound, in the absence of an auditory stimulus

Transport - to move across membranes

Trauma - a harmful event or the wound caused by such an event

U

Undershoot - the hyperpolarization phase of the action potential

Unidirectional flow - the movement of a substance across an interface in only one direction

V

Vectocardiography - the analysis of EKG signals, in which the signals are displayed as a spot moving in two or three directions.

Venesection - the removal of a pint of blood, in a way similar to that of donating blood, used to reduce the number of red cells or excess of iron

Vivisection - the practice of cutting a living organism, usually done on animals for scientific purposes

W

Wolffian duct - the region of the embryonic gonad that develops into the male reproductive ducts. This structure is missing from female individuals

X

Xerostomia - the scientific term for dry mouth

Z

Z disk (or **Z line**) - this structure defines the boundaries of the sarcomere. In other words, two adjacent **Z lines** encompass a single sarcomere.

CHAPTER 4

Skeletal Terms

The skeletal system has storage and protective functions. It protects the internal organs and serves as the primary storage site for calcium, phosphate, and other essential minerals. In addition, the bone marrow, found within the bones, houses tissues that produce blood cells while storing fat. The skeleton can be further subdivided into two divisions which are:

Axial skeleton

The Axial Skeleton forms the central vertical axis of the human body. It includes all the back, chest, neck, and head bones. It protects vital organs, including the lungs, heart, and brain. The axial skeleton also serves as the site of attachment for the muscles responsible for moving the back, neck, and head. In adults, the axial skeleton contains eighty bones, twenty-two of which form the skull.

Appendicular skeleton

The Appendicular Skeleton includes all the bones of the lower and upper limbs and the bones that connect each limb to the axial skeleton. The appendicular skeleton consists of one hundred and twenty-six bones.

WHAT IS THE SKELETAL SYSTEM?

The skeletal system is one of the body's major systems, which consists of bones, ligaments, and cartilage. It supports the body and gives shape to all its structures. Adults have two hundred and six bones in the skeletal system, while children have more because some of their bones haven't fused yet. The skeletal system's primary function is to give a firm internal structure to support the body's weight. It also provides the structure for which muscles can produce movements. The skeleton's lower part provides stability for actions such as running and walking. On the other hand, the skeleton's upper part has a greater range of motion and mobility to perform different functions.

The skeletal system has many bones and bony tissues that perform specific support and locomotion functions. For example, skeletal muscles move the bones while the skeletal system produces and stores blood cells in the bone marrow.

The parts of a bone are:

- **Diaphysis** which is the bone's long shaft.
- **Epiphysis,** which is the bone's end, is knob-like and contains blood cells (red marrow).
- **Metaphysis** which is essential in the growth of bones and where the diaphysis and epiphysis are joined.

- **Medullary** which is the marrow cavity within a bone that contains fat cells (yellow marrow).
- **Foramina** which are small canals within a bone where lymph and blood vessels are connected to the medullary.
- **Cartilage** which is found at the ends of a bone to connect it to other bones.
- **Ligament** which connects cartilage or bones while supporting and strengthening joints.

There are also different types of bones in the skeletal system which are:

- **Compact bones** which are hard and dense.
- **Spongy bones** which consist of latticework and have red marrow in the gaps.
- **Long bones** which are curved slightly making them stronger.
- **Short bones** which are usually spongy.
- **Flat bones** which are highly protective and have a plate-like structure.
- **Irregular bones** which have varying shapes.

Skeletal Terms

Now that you know more about the skeletal system, let's go through some of the most common skeletal terms:

Acetabulum - the cup-like hollow portion in the hip bone, in which the femur fits its head, forming a "ball-and-socket" joint (which has high mobility) The interior surface of the **acetabulum** is covered by a chitinous cuticle.

Acetabulum refers to the cup-shaped space in the pelvic bone connected to the femur's head.

Acetabulum is the socket in the bone of the hip which connects to the femur forming a ball-and-socket joint.

Acromion refers to the point of the spine or shoulder blade located at the outermost part.

Acromion refers to the outermost part of the spine or shoulder blade. The collar bone is connected to the **acromion**.

Acromion - the outermost portion of the shoulder blade

Ankle refers to the joint located between the fibula and tibia.

Ankle is the joint located between the fibula and tibia that is capable of gliding. Injuries to the **ankle** makes it difficult to walk.

Bone is a rigid type of tissue that's part of the skeleton.

Bone is a part of the skeleton that is a rigid type of tissue. Breaking a **bone** requires immediate treatment.

Calcaneus is the largest of the tarsal bones.

Calcaneus is the biggest tarsal bone which forms the heel. The **calcaneus** is one of the most common sites of stress fractures.

Cancellous refers to having a latticed, porous, or open structure.

Cancellous means having an open, porous or latticed structure. The tympanic bulla contains **cancellous** tissue.

cancellous - having a porous structure

Cartilage refers to a tough but elastic tissue, most of which gets converted to bone.

Cartilage is a whitish, elastic, tough tissue that gets converted to bone in adults. The very first rib is more slender, shorter, and is connected to costal **cartilage**.

Cavity refers to a space that's enclosed.

Cavity is a space that's surrounded or enclosed by something. Each **cavity** in the body protects the organs it houses.

Clavicle refers to the bone which links the sternum and scapula.

Clavicle is the bone between the sternum and scapula which links them. The **clavicle** is one of the bones most associated with scapula fractures.

Coccyx is the small bone located at the end of the vertebral column.

Coccyx refers to a bone with a triangular shape at the end of our vertebral column. During fetal development, the **coccyx** starts as a tail-like structure.

Condyle refers to the round bump on bones at the point where a joint forms with other bones.

Condyle is a bump on a bone with a round-shape where the bone forms a joint. A **condyle** looks like a rounded prominence at the ends of bones.

Diaphysis refers to the main section of long bones.

Diaphysis is the mid-section or shaft of long bones. All long bones have a **diaphysis**, distal epiphysis, and proximal epiphysis.

Epiphysis refers to the end of long bones.

Epiphysis refers to the end part of long bones. There are rare cases when osteomyelitis starts in the **epiphysis**.

Ethmoid refers to one of the eight cranial bones.

Ethmoid is one of the eight bones which make up the cranium. The **ethmoid** in mammals ossifies forming the turbinals.

Fibula refers to the thinner bone found between the ankle and knee.

Fibula refers to the outer, thinner bone of the lower leg. The **fibula** and tibia are inferiorly united.

Fissure refers to a narrow and long depression on surfaces.

Fissure is a depression on a surface that's narrow and long. A **fissure** may form because of an injury.

Fontanelle refers to the gap between the bones of t

he skull of an infant.

Fontanelle is the gap between the bones of the skull of an infant covered by a membrane. The **fontanelle** of infants is very delicate.

Foramen is a natural perforation or opening through a membranous structure.

Foramen refers to a natural perforation or opening through a membranous structure or bone. Intrajugular processes may divide the jugular **foramen** into two separate parts.

Humerus refers to the bone extending from the elbow to the shoulder.

Humerus is the long bone of the forelimb that extends from the elbow to the shoulder. A fracture of the **humerus** may be caused by a fall or a direct blow onto an outstretched arm.

Ilium is the upper bone of the bones that make up the hipbone.

Ilium refers to the widest of the bones that make up the hipbone located at the upper part. The hipbone consists of three bones namely the ischium, pubis, and **ilium**.

Intervertebral refers to the gap between two vertebrae.

Intervertebral refers to the space or gap between two vertebrae. **Intervertebral** discs must be able to withstand heavy compressive loads.

Ischium refers to one of the bones that make up the hipbone.

Ischium refers to one section of the hipbone. The **ischium** is located at the lowest part of the hipbone.

Joint refers to the point where bones are linked with each other.

Joint is the point of connection between bones. A swollen **joint** may cause pain and immobility.

Lamina refers to a thin layer of bone.

Lamina refers to a thin layer or plate of bone. A **lamina** is quite fragile because of how thin it is.

Ligament is a fibrous band of tissue connecting cartilages and bones.

Ligament refers to a band of fibrous tissue which connects cartilages and bones. A person who suffers from a torn **ligament** requires immediate treatment.

Manubrium refers to the upper section of the breastbone.

Manubrium is the breastbone's upper part. The **manubrium** is one of the three bones that make up the breastbone.

Medullary means consisting of bone marrow.

Medullary means containing bone marrow. At a **medullary** level, the pyramidal tract is dorsal to the medial lemniscus.

Metaphysis refers to the part of a long bone that grows between the epiphysis and diaphysis.

Metaphysis is a growing part of long bones located between the epiphysis and diaphysis. During development, the **metaphysis** keeps growing.

Olecranon is the process of the ulna forming the elbow's outer bump.

Olecranon is a large process of the ulna's upper portion that forms the elbow's outer bump. Patients who suffer from chronic gout are likely to have tophi over their **olecranon** prominence.

Osteoblast is a type of cell from which bones develop.

Osteoblast refers to a bone-forming cell. The exctoenzyme known as alkaline phosphatase is the key indicator of the maturation of an **osteoblast**.

Osteoclast is a type of cell with a function in breaking down and reabsorbing bone tissue.

Osteoclast refers to a cell which plays a role in breaking down and reabsorbing bone tissue. Isoflavonoid glabridin can potentially inhibit the maturation of **osteoclast.**

Patella is a small bone located in front of the knees.

Patella refers to a small, moveable bone with a triangular shape. A knee injury may also cause damage to the **patella**.

Pelvis a skeletal structure that supports the lower limbs.

Pelvis a structure supporting the limbs of the lower body. The aorta's parietal branches supply the walls of the **pelvis**.

Periosteum is a fibrous membrane which covers a bone's surface.

Periosteum is a dense and fibrous membrane which covers a bone's surface. The **periosteum** contains nerve fibers that transmit pain.

Pubis refers to the forward part of the hipbone.

Pubis is the hipbone's front arch. The obturator foramen is formed when the **pubis** joins with part of the ischium.

Rib refers to one of the curved bones which extends from the spinal cord to the sternum.

Rib is one of the twelve pairs of curved, arched bones. A broken **rib** can potentially harm the organs in the surrounding area.

Scapula is a flat bone located on the shoulder's side.

Scapula is a flat bone with a triangular shape on the side part of the shoulder. One function of the **scapula** involves protraction and retraction.

Sesamoid refers to a small bone with a round shape that forms in tendons.

Sesamoid is a small bone with a round shape that forms in tendons where they pass over joints. **Sesamoid** bones might develop in bicipital tendons.

Skull is the head's bony skeleton.

Skull is a bony structure at the top of the body. The **skull** is where the brain is located.

Sphenoid is a bone with a butterfly shape located at the bottom of the skull.

Sphenoid is a bone with a butterfly shape located at the skull's base. Orbitosphenoids form the **sphenoid's** lesser wings.

Sternum is also known as the breastbone.

Sternum is another word for the breastbone. There are a few "false rib" pairs that aren't connected to the **sternum**.

Talus is an ankle bone attached to the leg bones forming the joint of the ankle.

Talus refers to a bone found in the ankle attached to the leg bones to form the joint of the ankle. The **talus** is stabilized by the lateral and medial malleoli of the fibula and tibia.

Tibia refers to the thicker bone found between the ankle and knee.

Tibia refers to the inner, thicker bone of the lower leg. The fibula and **tibia** of the lower legs are fused together.

Trochanter is a bony prominence that develops near the femur's upper extremity.

Trochanter refers to a type of bony prominence which develops near the point where muscles are attached. The **trochanter** can act as a type of swivel joint for the legs.

Tuberosity refers to a bone protuberance for attachment of ligaments or muscles.

Tuberosity is a protuberance on bones, especially for ligament or muscle attachment. Sesmoid bones may also develop over radial **tuberosity.**

Ulna refers to the inner bone of the forearm.

Ulna is the inner longer bone of the forearm. The **ulna** articulated with carpal bones.

Vertebra is one of the spinal column's bony segments.

Vertebra refers to a bony segment in the spinal column. The basis of the classification of a **vertebra's** centrum depends on how its elements are fused.

Vertebral column is a series of vertebrae which keeps the spinal cord protected.

Vertebral column is a series of vertebrae which protects the spinal cord and forms the skeleton's axis. The **vertebral column** is part of the central nervous system

Vomer refers to a thin skull bone which forms the nasal septum's inferior and posterior parts.

Vomer refers to a thin skull bone with a trapezoidal shape which forms the nasal septum's inferior and posterior parts. The **vomer's** anterior ventral surface is ventrally convex and narrow.

ADDITIONAL COMMON SKELETAL TERMS TO REMEMBER

column - a line of units, following one after another, the spinal column

comminute - to be reduced to small pieces by abrading or pounding

compound - a structure formed by the union of two or more elements

compression - the act of applying pressure to something

fossa - a concavity in a bone structure

ilium bone - the uppermost and largest bone of the hip

impact - the packing of a body against another

lacrimal - relating to tears

SKELETAL TERMS

lesion - any abnormal structural change

radius - a straight line drawn from the center to the perimeter of a circle

sinus - an abnormal passage that connects a suppurating cavity to the surface of the body

spinous - shaped like a spine, or "pointy"

symphysis - a place in which two bones are joined, forming immovable joints or completely fusing together

talipes - the technical term for "club foot", where the foot is curled or twisted at the ankle

tibia - the inner, thicker, long bone of the calf

zygomatic - relating to the cheek (the prominence of the cheek) region of the face

CHAPTER 5

Prefixes and Suffixes Used in Anatomy and Physiology

Although Anatomy and Physiology are more focused on body-related topics, many terminologies are involved in these subjects. Therefore, it's important to know what they mean - which also applies to affixes. Prefixes and suffixes have the power to alter the meaning of a word, making it something entirely new. This chapter will review the most common affixes attached to medical root words. With this knowledge, you will gain a deeper understanding of medical terms whenever you encounter them.

COMMON PREFIXES

Now that you have a better understanding of how affixes work, let's go through the most common prefixes first:

(**Ana -**) can indicate an upwards motion or an increase in value. **Ana**goge, the spiritual interpretation of a text or a higher way of thinking.a build-up or synthesis (**ana**bolism). It may also indicate excess, separation or repetition (for example **ana**clisis means an excessive emotional attachment to others).

(**Angio-**) describes a type of cell or receptacle. **Angio**blast (the embryonic cell that develops into blood cells).

angioblastoma (tumors composed of angioblasts); **angio**edema (swelling of the deep layers of the skin)

(**Arthr-** or **Arthro-**) refers to a junction or a joint that separates different parts. Ex: **arthr**algia (pain of the joints).

arthrectomy (the surgical removal of a joint) and **arthro**logy (the branch of anatomy that focuses on the joints).

(**Auto-**) describes something as belonging to itself or as occurring spontaneously. **auto**antibodies (antibodies produced by an organism, that attacks its own cells).

autogenic (self-generating or something that comes from within). Example: **auto**lysis (the destruction of a cell by its own enzymes).

(**Cephal-** or **Cephalo-**) referring to the head. Example: **cephal**ic (something located near the head).

Cephalocele (a protrusion of the brain through an opening in the skull); **cephalo**pathy (any disease of the brain).

(**Chrom-** or **Chromo-**) denotes pigmentation. Example: **chrom**atic (referring to colors).

chromatophore (a cell that produces pigment). **chromo**gen (a substance that lacks color).

(**Diplo-**) means doubled or paired. Example: **diplo**bacilli (a bacteria that remains in pair after division).

Diplocardia (a heart condition in which the heart is split into two halves by a fissure). **Diplo**coria (a condition in which there are two pupils in one iris).

(**Ect-** or **Ecto-**) means outer or external. Ex: **ecto**cellular (something external to a cell); **ecto**cornea (the outer layer of the cornea); **ecto**cranial (a position external to the skull).

(**End-** or **Endo-**) means inner or internal. Example: **endo**crine (the secretion of a substance internally).

Endogenous (something produced within an organism); **endo**plasm (the inner portion of the cytoplasm).

(**Epi-**) indicates a position that is above or near a surface. Example: **epi**cardium (the outermost layer of the heart wall).

Epigastric (the upper middle region of the abdomen). **Epi**thelium (the layer that covers the outside of the body, or organs or cavities).

(**Erythr-** or **Erythro-**) something red or reddish in color. Example: **erythr**algia (pain or redness of the skin).

Erythrocyte (red blood cell); **erythro**psin (a visual condition that makes objects appear as reddish in color).

(**Ex-** or **Exo-**) means external or away. Example: **exo**derm (the outer germ layer of an embryo).

Exophoria (the tendency of an eye or both eyes to move forward). **Ex**coriation (a scratch on the outer layer of the skin).

(**Eu-**) means genuine or good. Example: **eu**diometer (an instrument that tests the quality of the air).

Euglena (single-cell protists that have a true nucleus); **eu**peptic (relating to good digestion).

(**Glyco-** or **Gluco-**) referring to sugar or sugar derivatives. Example: **gluco**meter (a medical device to measure blood **gluco**se concentration).

Glucophore (atoms that give a substance a sweet taste); **glyco**pexin (the storage of sugar in body tissues).

(**Haplo-**) means simple or single. Example: **haplo**id (a cell with a single set of chromosomes).

Haplopia (or single vision, where two objects seen by both eyes appear as a single object); **haplo**sis (the halving of chromosomes number during division).

(**Hem-** or **Hemo-** or **Hemato-**) relating to the blood or its components. Ex: **hemato**id (resembling blood).

Hematuria (the presence of blood in the urine). **Hemo**rrhage (abnormal flow of blood).

(**Heter-** or **Hetero-**) means different or unlike others. Example: **hetero**chromia (a condition in which the irises are of different colors).

Heterophil (having an affinity for a certain substance). **Hetero**psia (a condition in which someone has a different vision in each eye).

PREFIXES AND SUFFIXES USED IN ANATOMY AND PHYSIOLOGY

(**Karyo-** or **Caryo-**) refers to the nucleus of a cell, or to a nut or kernel. Example: **caryo**psis (a plant that has a single, seed-like fruit).

Karyogamy (the unity of cell nuclei, during fertilization). **Karyo**n (the nucleus of a cell).

(**Meso-**) means middle or intermediate. Example: **meso**colon (connects the colon to the abdominal wall).

Mesogastrium (the middle region of the abdomen). **Meso**cephalic (having a head size of medium proportions).

(**My-** or **Myo-**) related to the muscle. Example: **my**algia (muscle pain). **Myo**cele (muscle hernia); **myo**fibril (long thin muscle thread).

(**Neur-** or **Neuro-**) referring to nerves or the nervous system. Example: **neur**algia (acute pain extending along the nerves).

Neuritis (inflammation of a nerve); **neuro**logist (one who studies the nervous system).

(**Peri-**) meaning surrounding or around. Example: **peri**cardium (the membranous sac that protects the heart).

Periosteum (a dual-layered membrane that covers the exterior surface of the bones); **peri**tubular (something that's close to a tube structure).

(**Phag-** or **Phago-**) related to eating or swallowing. Example: **phago**cyte (a white blood cell that consumes waste material).

Phagophobia (an irrational fear of swallowing); **phago**lysis (the destruction of a phagocyte).

(Poli-) means many or excessive. Example: **poli**odystrophy (atrophy of cerebral gray matter).

Poliomyelitis (an inflammatory process involving the grey matter); **poli**osis (absence of pigment of the hair).

(Proto-) means primitive or primary. Example: **proto**blast (a cell in the early stages of development).

Prototroph (an organism that can acquire nourishment from inorganic sources); **proto**zoa (unicellular organisms considered to be the first "animals").

(Staphyl- or **Staphylo-)** referring to a bunch or a cluster. Example: **staphylo**coccus (parasite bacterium that forms grape-like clusters).

Staphyloma (a protrusion of the cornea caused by inflammation); **staphylo**ncus (swelling of the uvula).

(Tel- or **Telo-)** referring to an extremity, and end or a terminal phase. Example: **Telo**centric (a chromosome that has the centromere near its end region).

Telodendron (the terminal branches of the axon); **telo**mere (a protective cap located at the end of a chromosome).

(Zo- or **Zoo-)** referring to animals or animal life. Example: **zoo**biotic (a parasitic organism living in or on an animal).

Zooblast (an animal cell); **zoo**genesis (the origin and development of animal life).

MORE PREFIXES TO REMEMBER

- **a-** or **an-** which means a lack of
- **ab-** which means away from
- **acou-** which means hearing
- **acro-** which means extremity
- **acu-** which means needle
- **ad-** which means near to, toward
- **adeno-** which means gland
- **adipo-** which means fat
- **aero-** which means gas, air
- **allotri-** which means foreign, strange
- **ambi-** or **amphi-** which means both
- **ana-** which means an upward direction, buildup, synthesis, repetition, separation, excess
- **andro-** which means male
- **angio-** means a kind of receptacle
- **ante-** means forward, before
- **anti-** means reversed, against
- **apo-** means off, separation

- **arthr-** or **arthro-** means a junction or a joint separating various parts
- **atelo-** means incomplete
- **auto-** means occurring within, occurring spontaneously, belonging to oneself
- **basi-** means foundation, base
- **bi-** means double, twice
- **bio-** means live
- **blast-** means a developmental stage that's immature
- **brady-** means slow
- **brevi-** means short
- **bucco-** means cheek
- **carcin-** means cancer
- **cardio-** means heart
- **cata-** means lower, down
- **cephal-** or **cephalo-** means head
- **celio-** means abdomen
- **cerebro-** means brain
- **chol-** means bile
- **cholecyst-** means gallbladder
- **chondr-** means cartilage
- **chrom-** or **chromo-** means pigmentation, color
- **chrono-** means time

PREFIXES AND SUFFIXES USED IN ANATOMY AND PHYSIOLOGY

- **circum-** means about, around
- **co-**, **com-** or **con-** means together, with
- **colpo-** means vagina
- **contra-** means opposite, against
- **copro-** means feces
- **crypto-** means hidden
- **cyano-** means blue
- **cysto-** means sac, bladder
- **cyto-** or **cyte-** means cell
- **dactyl-** or **-dactyl** means tactile appendages, digits
- **de-** means away from
- **demi-** means half
- **derm-** means skin
- **di-** means two
- **dia-** means apart, through, across
- **diplo-** means paired, twofold, double
- **dis-** means apart from, reversal
- **dys-** means bad, difficult
- **e-** means away from, out
- **ec-** means out from
- **ect-** or **ecto-** means outer, external
- **em-** or **en-** means in

- **end-** or **endo-** means inner, internal, within

- **entero-** means intestine

- **epi-** means above, on, upon, near a surface

- **erythr-** or **erythro-** means red, reddish color

- **eu-** means true, genuine, good, well

- **ex-** or **exo-** means external, away from, out of

- **extra-** means outside

- **galacto-** means milk

- **gastro-** means stomach

- **gloss-** means tongue

- **gam-** or **gamo-** means fertilization, marriage, sexual reproduction

- **glyco-** or **gluco-** means sugar, sugar derivative, sweet

- **haplo-** means simple, single

- **helio-** means sun

- **hem-**, **hemo-** or **hemato-** means blood, blood components

- **hemi-** means half

- **hepato-** means liver

- **heter-** or **hetero-** means unlike, other, different

- **hexa-** means six

- **hist-** means tissue

- **homeo-** means same

- **hydro-** means water, wet
- **hyper-** means over, excessive, above
- **hypo-** means under, deficient, below
- **hyster-** means uterus
- **idio-** means self-produced
- **im-** means not
- **in-** means into, in
- **infra-** means below
- **inter-** means between
- **intra-** means within
- **ipsi-** means same
- **iso-** means equal
- **juxta-** means adjoining, near
- **karyo-** or **caryo-** means kernel, nut, nucleus of a cell
- **kerato-** means horny tissue, cornea
- **lact-** means milk
- **leuko-** means white
- **lipo-** means fat
- **litho-** means stone
- **macro-** means large
- **mal-** means bad
- **malaco-** means soft

- **mammo-** or **mast-** means breast
- **mega-** means great
- **melano-** means black
- **meso-** means middle, intermediate, mid
- **meta-** means beyond, change, after
- **micro-** means small
- **mito-** means filament, thread
- **mono-** means single, one
- **multi-** means much, many
- **myelo-** means spinal cord, marrow
- **my-** or **myo-** means muscle
- **narco-** means numbness
- **necro-** means death
- **neo-** means new
- **nephro-** means kidney
- **neur-** or **neuro-** means nerves, nervous system
- oculo- or ophthalm- means eye
- **odonto-** means tooth, teeth
- **oligo-** means few, little, scanty
- **omni-** means all
- **ortho-** means normal, straight
- **osteo-** means bone

PREFIXES AND SUFFIXES USED IN ANATOMY AND PHYSIOLOGY

- **oto-** means ear
- **pan-** means all
- **para-** means beyond, near to, beside
- **pedo-** means child
- **penta-** means five
- **per-** means excessive, through
- **peri-** means surrounding, around, near
- **phag-** or **phago-** means eating, consuming, swallowing
- **phleb-** means vein
- **phlogo-** means inflammation
- **phyto-** means plant
- **pleo-** means more than one
- **pneumo-** means lungs, air, gas
- **pod-** means foot
- **poli-** means many, excessive, much
- **post-** means behind, after
- **pre-** or **pro-** means in front of, before
- **procto-** means rectum, anus
- **proto-** means primary, or primitive
- **pseudo-** means false
- **psycho-** means soul, mind
- **pyo-** means pus

- **quasi-** means resembling, almost
- **re-** means again, contrary, back
- **reno-** means kidney
- **retro-** means located behind, backward
- **rhino-** means nose
- **sarco-** means flesh, fleshy
- **sclero-** means hard
- **semi-** means half
- **somato-** means body
- **staphyl-** or **staphylo-** means bunch, cluster
- **steno-** means narrow
- **sub-** means under
- **super-** or **supra-** means upper, excessive, above
- **sym-** or **syn-** means together, with
- **tachy-** means fast, swift
- **tel-** or **telo-** means denoting an end, final phase, extremity
- **tetra-** means four
- **therm-** means heat
- **toco-** means labor
- **tox-** means poison
- **trans-** means through, beyond, across

- **tri-** means three

- **ultra-** means beyond, excess

- **uni-** means one

- **vaso-** means vessel

- **vene-** means vein

- **viscer-** means internal organ

- **vivi-** means alive

- **xeno-** means foreign

COMMON SUFFIXES

(**-ase**) referring to an enzyme, added at the end of the substrate name. Example: amyl**ase**; collagen**ase**; histamin**ase**.

(**-ectomy** or **-stomy**) the act of surgical removal or cutting. Example: append**ectomy** (removal of the appendix).

embol**ectomy** (removal of an embolus); ileo**stomy** (creating an opening in the abdomen to reach the illeon).

(**-emia** or **-aemia**) referring to a blood condition or the presence of something in the blood. Example: an**emia** (low red cell count).

leuk**emia** (cancer of the white blood cells); hypocalc**emia** (low levels of calcium in the blood).

(**-genic**) means producing or forming. Ex: terato**genic** (something that causes physical defects in an embryo).

Aorto**genic** (something that starts in the aorta); karyo**genic** (relating to karyogenesis).

(**-itis**) referring to inflammation of organs or tissues. Example: alveol**itis** (inflammation of alveoli); capillar**itis** (inflammation of capillaries).

Epicard**itis** (inflammation of the epicardium).

(**-kinesis** or **-kinesia**) indicating movement or activity. Example: cyto**kinesis** (a process in which the cytoplasm of a cell divides).

Nucleo**kinesis** (the forward movement of a nucleus). Electro**kinesis** (the movement of transport particles due to the action of an electric field).

(**-lysis**) referring to decomposition, degradation or releasing. Ex: auto**lysis** (self-destruction of tissues); chemo**lysis** (decomposition of organic substances); dia**lysis** (separation of smaller molecules from larger molecules in a solution).

(**-oma**) referring to an abnormal growth such as a tumor. Ex: melan**oma** (skin cancer); granul**oma** (abnormal structure formed by immune cells); sarc**oma** (a malignant tumor of connective tissues).

(**-osis** or **-otic**) referring to a disease or abnormal production of substances. Example: cirrh**osis** (chronic liver disease).

Thromb**osis** (a condition that involves the formation of blood clots); cyan**otic** (relating to cyanosis, where the skin appears blue due to lack of oxygen).

(**-otomy** or **-tomy**) referring to an incision or surgical cut. Example: gastr**otomy** (a surgical incision done into the stomach.

Lapar**otomy** (incision done in the abdominal wall to expose the organs); lob**otomy** (incision done into a lobe of a gland or organ).

(**-penia**) relating to a deficiency or a lack of something. Example: calci**penia** (insufficient amount of calcium in the body).

Cyto**penia** (insufficient production of a certain blood cell); glyco**penia** (sugar deficiency in an organ or tissue).

(**-phage** or **-phagia**) the act of eating. Ex: aero**phagia** (the act of swallowing excessive amounts of air); a**phagia** (the inability to swallow); macro**phage** (a white blood cell)

(**-phile** or **-philic**) having a strong attraction or affinity towards something. Ex: acido**phile** (organisms that thrive in acidic environments); halo**phile** (organisms that thrive in environments with high concentrations of salt); auto**philia** (a narcissistic type of self-love).

(**-plasm** or **-plasmo**) referring to tissue or a living substance. Example: axo**plasm** (the cytoplasm of a nerve cell).

Myo**plasm** (the portion of a muscle cell that contracts); neo**plasm** (abnormal growth of new tissue).

(**-scope**) referring to an instrument used for observation. Ex: angio**scope** (microscope used to examine capillary vessels).

Baro**scope** (an instrument that measures atmospheric pressure); endo**scope** (an instrument for examining internal body caviti).

(-**stasis**) indicating the maintenance of a constant state. Example: apo**stasis** (the end stage of a disease).

Entero**stasis** (the slowing down of matter in the intestines); fungi**stasis** (the slowing down of fungal growth).

(-**troph** or -**trophy**) referring to nourishment or acquisition of nutrients. Ex: auto**troph** (an organism that self-nourishes).

Bio**troph** (parasites that establish a long-term infection); chemo**trophy** (an organism that makes its own energy through oxidation).

Examples are the best to give you a clear, well-rounded idea of how an affix can change the meaning of a word base. Try breaking words apart, on paper, to better understand how the word came to be. Once you get used to how the words sound along with their meaning, you'll have an easier time remembering them.

ADDITIONAL SUFFIXES

Now let's go through the most common suffixes used in Anatomy and Physiology:

- **-able** means capable
- **-agogue** means inducing, leading
- **-al** expressing relationship
- **-algia** means pain
- **-ary** means associated with
- **-ase** means enzyme

114

PREFIXES AND SUFFIXES USED IN ANATOMY AND PHYSIOLOGY

- **-asis** means state of, condition
- **-c** expressing relationship
- **-cele** means cavity, hollow
- **-cide** means kill
- **-derm** or **-dermis** means tissue, skin
- **-dynia** means pain
- ~~**-ectomy** or **-stomy**~~ means cutting out, surgical removal
- **-emia** or **-aemia** means blood, a condition of the blood, some kind of substance present in the blood
- **-facient** means making, causing
- **-ferent** means carry
- **-ferous** means bearing, producing
- **-form** means expressing resemblance
- ~~**-gen**~~ means producing
- **-genesis** means origin, produce
- **-genic** means producing, forming, giving rise to
- **-gram** means a drawing
- **-graph** means an instrument which records
- **-ia** or **-id** expressing condition
- **-ism** means state of, condition
- **-itis** means inflammation
- **-ity** expressing condition

- **-kinesis** or **-kinesia** means movement, activity
- **-logy** means study
- **-lysis** means dissolving, degradation, bursting, releasing, decomposition, breaking up
- **-mania** means madness
- **-oid** expressing resemblance
- **-oma** means tumor, abnormal growth
- **-ory** means referring to
- **-ose** means full of
- **-osis** or **-otic** means a disease, abnormal substance production
- **-otomy** or **-tomy** means an incision, surgical cut
- **-ous** expressing material
- **-pathy** means disease
- **-penia** means a lack, deficiency
- **-phage** or **-phagia** means eating, consuming
- **-phile** or **-philic** means affinity for, strong attraction
- **-phobia** means fear
- **-plegia** means paralyze
- **-plasm** or **-plasmo** means tissue, living substance
- **-rrhagia** means pour, burst forth
- **-rrhea** means discharge, flow

- **-scope** means an instrument for examination, observation

- **-stasis** means maintaining a constant state

- **-stat** means inhibit

- **-stomy** means make an artificial opening

- **-tomy** means incise, cut

- **-tropic** means influencing, changing

- **-uria** means urine

By this point, you're probably on board with the fact that anatomy and physiology terms are not the easiest to comprehend. But, here's the good news: a lot of them make use of affixes, making them easier to identify and memorize. What are affixes? Elements that are added before or after the base of a word. If the added part is before the base, then we call it a prefix, and if it is after the base, we call it a suffix. By knowing what these additional groups of words mean, complex terms will suddenly become easier to comprehend.

Affixes that act as both prefixes and suffixes

Blast- means a developmental stage that's immature (**Blast-** and **-blast**) shows an immature developmental stage. Example: **blast**ocyst (the developing fertilized egg).

Blastoma (a type of cancer developed in blast cells); **blast**oderm (the outer layer of the blastocyte); megalo**blast** (a large erythro**blast**); epi**blast** (the outer layer of the blastula).

Cyto- or **cyte-** means cell (**Cyto-** and **-cyte**) something that relates to a cell. Example: **cyto**plasm (the contents of a cell, excluding the nucleus).

Cytotoxic (a substance or process that kills cells); **cyto**kinesis (the division of the cell); adipo**cyte** (fat cell); erythro**cyte** (red blood cell); granulo**cyte** (a type of white blood cell).

(**Dactyl-** and **-dactyl**) refers to a digit or tactile extremities such as a finger or toe. Example: **dactyl**odynia (pain in the fingers).

dactyloid (the shape of a finger); **dactyl**olysis (a loss of a finger); a**dactyl**y (the absence of fingers/toes at birth).

Di**dactyl** (an organism with only two fingers per hand or toes per foot); mono**dactyl** (an organism with only one finger per hand or toe per foot).

(**Derm-** or **-derm/-dermis**) referring to skin or tissue. Example: **derm**atitis (skin disease); **derm**atology;

Dermatologist; epi**dermis**; ecto**derm** (outer layer of an embryo that develops skin); hypo**dermis**.

(**Gam-** and **-gamy**) refers to sexual reproduction or marriage. Example: **gam**omania (an obsessive desire to marry);

Gamic (having a sexual character); **gam**ogenesis (reproduction); hiero**gamy** (sacred marriage); karyo**gamy** (a fusion of cell nuclei occurring in fertilization); pseudo**gamy** (false marriage).

REVIEW

To avoid confusion and increase precision, anatomists and health care specialists use terminology when they refer to the human body structure. A thorough terminology study leaves no place for ambiguities, making it faster and easier for specialists to assess medical situations. However, most anatomical terms are derived from ancient Latin and Greek words, making them harder to memorize and understand at first glance. Why were these "dead languages" chosen as the basis for the medical lexicon? Specifically, because they are not in current use, ensuring that the meaning of the words won't change and that all anatomists can understand the anatomical terminology regardless of their native tongue.

ANATOMICAL POSITION

Since different angles and points of view of the human body can describe a human body, a standard position had to be agreed upon. So, all anatomists and health care specialists could find their way about it. A body that is in anatomical position is upright, facing the observer. The feet are flat and directed forward, while the upper limbs are at the body's sides, with palms facing forward. When talking about the placement of the internal organs, such as the liver, it's always positioned on the right side while the stomach is on the left. Therefore, there is no ambiguity regarding what is where. Whenever you are having difficulties describing where a body part, organ, or wound is, always picture the anatomical position with your mind's eye, and go from there.

Regarding the position of a human body lying down, there are two terms to keep in mind, as they are used when describing examinations or procedures. The term supine refers to a body that lies down, with the face oriented upwards, and prone relates to a body that lies down, with the face down. The most frequently used positions for medical examinations are in supine orientation, with the face upwards.

Some examples are the horizontal recumbent position (with the patient on their back, legs extended, and arms placed in various positions such as above the head, on the chest, or alongside the body).

Fowler's position (with the upper body elevated and the knees slightly raised, used when the patient has trouble breathing or when there is a need for drainage).

Dorsal Lithotomy Positions (seen primarily on gynecological examinations, with the patient lying down and the legs separated and placed in stirrups).

Prone orientation is usually used for medical exams concerning the spine or regions of the back, kidneys, rectum, etc. Some examples are the Prone Position (patient lying flat on their stomach, with the head turned to one side) and Sim's Position (patient lying on the left side, with the right knee flexed against the abdomen, used for rectal exams).

REGIONAL TERMS

Multiple regions make up the human body, each with a particular term to increase precision. Starting from top to bottom, the main regions of the human body are the:

- Cephalic (the head).
- Cervical (the neck); thorax/thoracic/thoracis (the chest).
- Abdominal (the abdomen).
- Pelvic (the pelvis region).
- The upper extremity (the upper limb in its entirety).
- Lower extremity (the lower limb in its entirety).

Because everything was decided with precision in mind, anatomists made things easier for themselves by taking these general regions and deciding upon subdivisions.

The Cephalic Region

1. cranial region/**cranium** (the skull): made of a frontal region/ **frons** (the forehead); a temporal region (the temples) and an occipital region (the base of the skull or the nape/back of the head)

2. facial region/**facies** (the face): made of the nasal/**nasus** region (nose); oral region/**oris** (mouth); orbital/ocular

region/**oculus** (eyes); buccal/**bucca** region (the cheeks); otic/**auris** region (ear); and the mental region (chin)

As you may notice, multiple terms refer to the same anatomical region. All of them are correct, and you can use whichever variation you find easier to memorize. However, I highly recommend learning the Latin version (the ones written in Bold) whenever you are in doubt.

The cervical region has no subdivision and is taken along with the cephalic region or the head and neck. It encompasses the Cervical portion of the spinal column, made of seven vertebras.

The Thorax

- the axillary region/ **axilla** (the armpit)
- the costal region (the ribs)
- the deltoid (the shoulder)
- the scapular region (shoulder blades)
- the pectoral region (the chest)
- the mammary region/**mamma** (the breasts)
- the sternal region (the breastbone)
- the vertebral region (backbone), which in this area is made of 12 vertebras and called Thoracic

The Abdomen

- the abdominal region
- the **umbilicus**/umbilical region (the navel)

- the Lumbar portion of the spinal column, made of five vertebras
- the pelvic region (the area that is between the hip bones)- although the pelvis could be taken on its own as a region it's usually "merged" in with the abdomen
- the **inguen**/inguinal region (the bend of the hip or the groin)
- the pubic region/**pubis** (the area surrounding the genitals)
- the gluteal region/**gluteus** (the buttocks)
- the perineal region (the area between the external genitalia and the anus)
- the Sacral (five vertebras) and Coccygeal (tailbone) sections of the spinal column.

The Upper Extremity

- the brachial region/ **brachium** or simply the arm (the upper arm)
- the cubital region with the antecubital region/ **antecubitis** (the inner elbow) and the olecranal region/**olecranon** (the back of the elbow)
- the antebrachial region/ **antebrachium** or simply forearm (the lower arm)
- the carpal region/ **carpus** (the wrist)
- the manual region/**manus** (the back of the hand)
- the palmar region/ **palma** (the palm of the hand)

- the digital region/ phalanges/**digits** (the fingers); out of the fingers the thumb is referred to as the **Pollex**

The Lower Extremity

- the femoral region/ **femur** or simply the thigh (the upper leg)
- the patellar region/ **patella** (the front of the knee)
- the popliteal region/ **popliteus** (the back of the knee)
- the crural region/ **crus** (the shin, the front of the lower leg)
- the sural region/ **sura** (the calf, the back of the lower leg)
- the tarsal region/ **tarsus** (the ankle)
- the pedal region/ **pes** (the foot)
- the plantar region/**planta** (the sole of the foot)
- the calcaneal region/ **calcaneus** (the heel of the foot)
- the digital region/ phalanges/ **digits** (the toes); out of which the great toe is referred to as the **Hallux**

The best way to learn each region is by using a body figure or even your own body in front of a mirror. While going through them, associate each term with its designated area. Try to refer to your body parts by their anatomical name instead of the usual terminology to get used to it better. Sure, it may sound weird at first to say "brachium" instead of the upper arm or "patella" instead of the knee, but some, such as the "Palma," are not that far from their familiar name, ex: palm.

DIRECTIONAL TERMS

When describing the location of an anatomical part, the more precise the terms are, the better. Directional terms are crucial when locating any body structure, so memorizing them and understanding their meaning is very important.

- Anterior (or **ventral**) describes the front of the body or something that is towards the front. For example, the nose is anterior to the face; the toes are anterior to the foot; the kneecap is on the anterior side of the leg.

- Posterior (or **dorsal**) describes the back of the body or something that is towards the back. For example, the shoulder blades are on the posterior side of the body; the **olecranon** is posterior to the **antecubitis**; the popliteus is posterior to the patella.

- Superior (or **cranial**) describes something that is located toward the head end of the body or that is above or higher than another part of the body. For example, the nose is superior to the mental region; the hand is part of the superior extremity; the pelvis is superior to the thigh.

- Inferior (or **caudal**) describes something that is further away from the head end of the body and closer to the lowest part of the spinal column, or that is situated below another body part. For example, the foot is part of the inferior extremity; the hand is inferior to the

antebrachium; the umbilic is inferior to the pectoral region.

- Lateral describes something located to the side of the body, away from the midline of the body. For example, the thumb (Pollex) is at the lateral side of the hand; the ears are lateral to the eyes; the eyes are lateral to the nose.

- Medial describes something to the midline of the body (the imaginary line that divides the body into two equal right and left halves). For example, the nose is medial to the eyes; the **hallux** is the medial toe; the sternal region is medial to the axilla.

- Proximal describes something that is closer to the trunk or the point of origin/attachment of a part. For example, the **brachium** is proximal to the **antebrachium**; the kneecap is proximal to the foot; the proximal end of the femur joins with the pelvic bone.

- Distal describes something that is further away from the trunk or from its point of attachment. For example, the **crus** are distal to the **femur**; the **carpus** is distal to the **antebrachium**; the **digits** are distal from the **carpus**.

- Superficial describes something that is situated near the surface of the body. For example, the skin is superficial to the skeletal muscles; the nose is superficial to the nasal bones; the epidermis is superficial to the hypodermis.

- Deep describes a position that is closer to the interior of the body and farther away from the surface. For example, bones are deeper than muscles; the lungs are deeper to the skin; the heart is deeper to the sternum.

BODY PLANES

Body planes, also known as reference planes, are imaginary geometric planes that divide the body into different sections. These imaginary planes are two-dimensional structures that pass through the body, showing a specific area. Because medical imaging devices display virtual sections of a living body, you must know the plane along which the section is made.

Three principal planes that are frequently used in anatomical terminology:

- The sagittal plane (or lateral plane) divides the body or a part of the body that is studied vertically, into a right and left side. If the sagittal plane is in the midline through the center of the body, it's called the midsagittal plane (or median). This midsagittal plane passes through the navel or the spine, and all the other sagittal planes are parallel to it.

- The frontal plane (or coronal) divides the body or a part of the body, into a ventral (front) and dorsal (back) portion. This plane is perpendicular to the ground, and it separates the structure that it goes through into its anterior and posterior sides.

- The transverse plane (or axial) divides the body or a part of the body into a superior and inferior portion. This plane is parallel to the ground, and the images produced by referring to this plane are called cross-sections.

Although these planes are hypothetical, they are arbitrarily chosen, and we can't deny that they allow anatomists and specialists to describe the location of a structure or the direction of movement. But they are also very tricky to imagine with your mind's eye. The best way to remember them is by studying simplified figures showing how these planes go through the human body and by looking at some scans made through each section. Choose an organ or area of the body (such as the thorax) and see how they look from a frontal section, transversal section, and sagittal section. It will give you a better understanding of the matter.

BODY CAVITIES

By this point, you know how to describe the placement of a body part or organ using regional terms, directional terms, and referring to body planes. But, when it comes to internal organs, these terms alone are not enough. That's why anatomists and specialists divided the body's insides into specific regions or cavities.

The human body has two main cavities, the ventral cavity (or anterior), the larger of the two and has divisions of its own, and the dorsal cavity (or posterior).

REVIEW

The ventral (anterior) cavity is formed by the thoracic cavity and the abdominopelvic cavity, separated by the diaphragm.

The thoracic cavity (or the upper ventral cavity) contains important organs (heart, lungs, trachea, esophagus), large blood vessels, and nerves and is enclosed by the rib cage.

The abdominopelvic cavity has two subdivisions: the abdominal and pelvic portions. The abdominal cavity contains most of the gastrointestinal tract, the kidneys, and the adrenal glands. Also, this abdominal cavity has nine regions to make placing different organs more precise. Taking them from the right side to the left and from superior to inferior, the regions are right hypochondriac; epigastric; left hypochondriac; right lumbar; umbilical; left lumbar; right iliac; hypogastric; left iliac. The sides are easy to remember because the name stays the same, they are just the right or the left division, and the umbilical one corresponds to the position of the umbilic. The pelvic bones bound the pelvic cavity, which encompasses most of the urogenital system and the rectum.

The dorsal cavity contains the organs that reside on the posterior side of the body. It has an upper portion, named the cranial cavity, which contains the brain, and a lower portion, named the vertebral canal, which houses the spinal cord.

Q & A

Q: What term can you use to describe a body that is lying with the face oriented upwards?

A: Supine

Q: How is a body positioned for a spinal exam?

A: In the Prone Position.

Q: What term describes the area above the elbow?

A: The arm or the brachium.

Q: What are the two regions of the hand?

A: The manus and the Palma.

Q: Is there a different terminology for the fingers and the toes?

A: No, they are both called digits or phalanges.

Q: What are the two regions of the lower leg?

A: The crus or the shin, and the sura or the calf.

Q: What term is more anatomically accurate: upper leg or thigh?

A: Thigh.

Q: Name the anatomical regions of the foot.

A: The pedal region, the plantar region, the calcaneal region, and the digital region.

Q: What are the two components of the cephalic region?

A: Cranium and facies.

Q: The shoulder and the axilla are part of the: thorax region or the upper extremity region?

A: The thorax.

Q: What is the term "inguen" referring to?

A: The inguinal region or the groin.

Q: What is the term auris referring to?

A: The otic region or the ear.

Q: What is the direction towards the front of the body called?

A: Anterior (or ventral).

Q: How is the spinal column situated in relation to the sternum?

A: The spinal column is posterior (or dorsal) to the sternum.

Q: How do you describe something that is located towards the head end of the body?

A: As being superior (or cranial).

Q: How is the abdomen situated in relation to the thorax?

A: The abdomen is inferior (or caudal) to the thorax.

Q & A

Q: What is the finger that is situated to the most lateral part of the hand?

A: The thumb (Pollex)

Q: Which is more medial: the nose or the ears?

A: The nose.

Q: What is the proximal end of the brachium?

A: The point at which it meets the shoulder joint.

Q: How are the digits situated in relation to the Tarsus?

A: The digits are distal to the Tarsus.

Q: Is the epidermis the most superficial layer of the skin?

A: Yes, it is, as it's the surface layer.

Q: What can be described as being deeper, the lungs or the ribs?

A: The lungs are deeper.

Q: How is the plane that divides the body into a right and a left side called?

A: The sagittal plane (or lateral).

Q: What is the plane that divides the body into its anterior and posterior portions?

A: The frontal plane (or coronal).

Q: What plane produces sections referred to as cross-section in medical imaging exams?

A: The transverse plane (or axial).

Q: Which are the two main cavities of the body?

A: Ventral (anterior) and dorsal (posterior).

Q: The ventral cavity has how many divisions? What are they called?

A: The ventral cavity has two main divisions: the thoracic cavity, and the abdominopelvic cavity.

Q: How many divisions does the abdominal cavity have? Name them.

A: Nine divisions and they are called: right hypochondriac; epigastric; left hypochondriac; right lumbar; umbilical; left lumbar; right iliac; hypogastric; left iliac.

Q: What is the upper portion of the dorsal cavity called and what does it contain?

A: It's called the cranial cavity and it contains the brain.

Q: What does the vertebral canal contain?

A: The spinal cord.

CONCLUSION

There you have it! All the essential information and terminology you need to ace your exams or participate in conversations with other medical professionals in the field. We have gone through some important concepts and terminology about Anatomy and Physiology.

The first time you read this study guide, you probably breezed through it to get a better idea of everything it covers. That's a great start! But when it's time for you to focus on learning, understanding, or reviewing the concepts and terms, it would be beneficial to write down the main words on a separate sheet of paper. After writing the main words down, you can even point to the actual body part it refers to. This guide provides examples for you wherein the terms are used in sentences. This guide helps you understand these terms better. Take it a step further by trying to develop your sentences using the necessary terminology. You can even ask a partner to help you memorize the terms as you go through this simple yet comprehensive study guide.

Now that you have reached the end of this book, this doesn't mean your journey is over. Remember, whether you purchase the audio version, the eBook version, or the physical version of the book, you can bring it wherever you go. That way, you can whip it out whenever you need a refresher or have free time to review concepts and terms for your upcoming exam. The bottom line is this: you now have a convenient, complete, and effective tool at your disposal to help you learn, remember, and understand all that you need about Anatomy and Physiology!

REFERENCES

1.6 Anatomical Terminology – Anatomy and Physiology.
(2019). Retrieved from
https://opentextbc.ca/anatomyandphysiology/chapter
/1-6-anatomical-terminology/

7.1 Divisions of the Skeletal System – Anatomy and Physiology.
(2019). Retrieved from
https://opentextbc.ca/anatomyandphysiology/chapter
/7-1-divisions-of-the-skeletal-system/

Allrich, R. Physiology Prefixes and Suffixes. (2005). Retrieved
from
https://web.ics.purdue.edu/~rallrich/space/prefix.htm
l

Anatomy & Physiology Study Tips. (2019). Retrieved from
https://www.registerednursern.com/anatomy-
physiology-study-tips/

Anatomy Study Guide for Health Students. (2019). Retrieved
from https://www.kenhub.com/en/help/study-
guides/anatomy

Bailey, R. Can You Speak Biology? Learn Common Prefixes and
Suffixes. (2019). Retrieved from

https://www.thoughtco.com/biology-prefixes-and-suffixes-373621

Bailey, R. Tips and Advice for Studying Human Anatomy. (2019). Retrieved from https://www.thoughtco.com/anatomy-s2-373478

Henderson, B., & Dorsey, J. Medical Terminology for Your Physiology. (2019). Retrieved 17 July 2019, from https://www.dummies.com/careers/medical-careers/medical-terminology/medical-terminology-for-your-physiology/

Making And Using Study Guides-Aids To Preparing For An Exam. (2019). Retrieved from https://www.purdue.edu/asc/resources/pdfs/study_guides.pdf

Newman, T. Introduction to physiology: History, biological systems, and branches. (2017). Retrieved from https://www.medicalnewstoday.com/articles/248791.php

Quia - Glossary of Physiology Terms. (2019). Retrieved from https://www.quia.com/jg/426788list.html

Rhodes, C. The Benefits of Study Guides: How a Simple Notebook Can Help You Make A's. (2012). Retrieved from https://ezinearticles.com/?The-Benefits-of-Study-Guides:-How-a-Simple-Notebook-Can-Help-You-Make-As&id=7012884

REFERENCES

Skeletal System - Vocabulary List. (2009). Retrieved from
    https://www.vocabulary.com/lists/22777

Smith, C. Anatomy and Physiology Vocab: Medical Suffixes.
    (2014). Retrieved from
    https://www.visiblebody.com/blog/anatomy-and-
    physiology-vocab-medical-suffixes

The Skeletal System (Bones). (2013). Retrieved from
    http://www.cancerindex.org/medterm/medtm6.htm

What is Physiology? - Definition & History. (2019). Retrieved
    from https://study.com/academy/lesson/what-is-
    physiology-definition-history.html

What makes physiology hard for students to learn?. (2007).
    Retrieved from
    https://www.physiology.org/doi/full/10.1152/advan.0
    0057.2006

1.6 Anatomical Terminology – Anatomy and Physiology.
    (2019). Retrieved 13 September 2019, from
    https://opentextbc.ca/anatomyandphysiology/chapter
    /1-6-anatomical-terminology/

Anatomical Terminology | SEER Training. (2019). Retrieved
    13 September 2019, from
    https://training.seer.cancer.gov/anatomy/body/termin
    ology.html

Smith, C. (2019). Anatomy and Physiology: Anatomical
    Position and Directional Terms. Retrieved 13

September 2019, from
https://www.visiblebody.com/blog/anatomy-and-physiology-anatomical-position-and-directional-terms

Rego, L. (2019). *Body Positions For Physical Exam* [Ebook]. Hawaii Tumor Registry.

Mapping the Body | Boundless Anatomy and Physiology. (2019). Retrieved 13 September 2019, from https://courses.lumenlearning.com/boundless-ap/chapter/mapping-the-body/

Norris, M., & Siegfried, D. (2019). The Anatomical Regions of the Body - dummies. Retrieved 13 September 2019, from https://www.dummies.com/education/science/anatomy/anatomical-regions-body/

Ebneshahidi, D. (2019). *Anatomical Terminology*. Presentation.

Gardner- Medwin, A., Curtin, N., & Tatham, P. (2019). PHYSIOLOGY GLOSSARY. Retrieved 13 September 2019, from https://www.ucl.ac.uk/lapt/glossph.htm

Team, P. (2019). Glossary of Physiology Terms and Phrases for Letter A - PhysiologyWeb. Retrieved 13 September 2019, from https://www.physiologyweb.com/glossary/a/index.html

Anatomy & Physiology Terms | Biology Dictionary. (2019). Retrieved 13 September 2019, from

REFERENCES

https://biologydictionary.net/category/anatomy-physiology/

What is physiology? - The Physiological Society. (2019). Retrieved 13 September 2019, from https://www.physoc.org/explore-physiology/what-is-physiology/

Bailey, R. (2019). Can You Speak Biology? Learn Common Prefixes and Suffixes. Retrieved 21 April 2019, from https://www.thoughtco.com/biology-prefixes-and-suffixes-373621

Cotterill, S. (1996). The Skeletal System (Bones) | Medical Terminology for Cancer. Retrieved 13 February 2014, from http://www.cancerindex.org/medterm/medtm6.htm

Skeletal System - Vocabulary List : Vocabulary.com. (2009). Retrieved 13 September 2019, from https://www.vocabulary.com/lists/22777

6.1 The Functions of the Skeletal System – Anatomy and Physiology. (2019). Retrieved 13 September 2019, from https://opentextbc.ca/anatomyandphysiology/chapter/6-1-the-functions-of-the-skeletal-system/

www.ingramcontent.com/pod-product-compliance
Lightning Source LLC
Chambersburg PA
CBHW031858200326
41597CB00012B/465